AF885056

SOTHIS PRESS

Sothis Press aims to completely reimagine the public domain book with an emphasis on universal access to captivatingly designed print editions. We believe that technology will enable every single book ever published to be perpetually accessible in beautiful print form. Sothis is beauty in ubiquity.

www.sothis.press
1 Waterhouse Square, London, ECIN 2ST

Design of Roof Trusses and Mill Buildings

Design of Roof Trusses Mill Design

I. C. S. Staff

SOTHIS PRESS

ISBN 978-1-0365-1467-9

Design of Roof Trusses and Mill Buildings
is in the public domain.

This edition is published by
Sothis Press © 2023. The company
is registered in England and Wales.
The Company Number is 14900389.

Design of Roof Trusses and Mill Buildings

By

I.C.S. STAFF
International Correspondence Schools

DESIGN OF ROOF TRUSSES
MILL DESIGN

273

Published by

INTERNATIONAL TEXTBOOK COMPANY

SCRANTON, PA.

Design of Roof Trusses: Copyright, 1907, by INTERNATIONAL TEXTBOOK COMPANY.
Mill Design: Copyright, 1907, by INTERNATIONAL TEXTBOOK COMPANY.

Entered at Stationers' Hall, London

All rights reserved

Printed in U. S. A.

INTERNATIONAL TEXTBOOK PRESS
Scranton, Pa.

76717

ummu
gift
9/12/02

CONTENTS

NOTE.—This book is made up of separate parts, or sections, as indicated by their titles, and the page numbers of each usually begin with 1. In this list of contents the titles of the parts are given in the order in which they appear in the book, and under each title is a full synopsis of the subjects treated.

DESIGN OF ROOF TRUSSES

MILL DESIGN

CONTENTS

MILL DESIGN—(*Continued*)

DESIGN OF ROOF TRUSSES

Serial 1095 —— Edition 1

LOADS ON ROOF TRUSSES

LIVE AND DEAD LOADS

1. Live Loads.—The loads on a roof·truss may be divided into four classes, namely, live, dead, snow, and wind. The **live loads** on roof trusses differ greatly in character. The load may be due to a hoisting crane or to a line of shafting attached to the chords; in some instances, also, roofs are used for summer gardens, gymnasiums, and the like. As a rule, however, a roof has no live load and the designer may omit it entirely in his calculations.

2. Dead Loads.—In obtaining the **dead load** on roof trusses, it is necessary, after having found the weight of the sheathing and roofing, to add a certain weight per square foot, to represent the weight of the truss or members supporting the roof. Not knowing, as yet, the size and weight of the different members in the roof truss, approximate weights must be assumed.

Table I gives the approximate weights of the trusses, or principals, as they are called, for roofs of different spans. These weights are, of course, only assumed, and may not be within 25 per cent. of the actual weight of the principals. They are, however, generally on the side of safety.

It is required, in the application of Table I, to obtain the weight in pounds per square foot of *roof surface*. As the

COPYRIGHTED BY INTERNATIONAL TEXTBOOK COMPANY. ALL RIGHTS RESERVED weights given in the table are in pounds per square foot of *area covered*, and as the area of the roof is considerably greater than this, owing to the pitch of the roof, it is necessary to divide the area covered by the area of the roof and multiply the result by the quantities given in the table. For example, the area of a building covered by a roof with a

TABLE I

POUNDS TO BE ADDED FOR WEIGHT OF PRINCIPALS, OR ROOF TRUSSES

Length of Span Feet	Weight per Square Foot of Area Covered Pounds
Up to 40	4
40 to 60	5
60 to 80	6
80 to 100	7

span of 50 feet is 10,000 square feet, and the area of the roof is 15,000 square feet; $10,000 \div 15,000 = \frac{2}{3}$, or .67, and, since the span of the roof is 50 feet, according to Table I, the weight of the truss is 5 pounds for each square foot of area covered. Therefore, $5 \times .67 = 3.35$ pounds must be added to the weight of each square foot of roof surface.

EXAMPLE.—The span of a roof truss is 40 feet, and its rise is 10 feet; what weight per square foot of roof surface should be assumed so as to allow for the weight of the principal or roof truss?

SOLUTION.—The area covered is 40 ft. per ft. in length of the building, while the area of the roof for the same length by the principles in geometry is $2\sqrt{10^2 + 20^2}$ or $2\sqrt{500}$. From Table I, the weight per square foot of area covered is 4 lb. for a span of 40 ft. Therefore, the weight per square foot of roof area is

$$4 \times \frac{40}{2\sqrt{500}} = 3.58 \text{ lb. Ans.}$$

SNOW AND WIND LOADS

3. Snow Loads.—In calculating the weight on roofs, there are two other loads always to be considered when obtaining the stresses on the various members of the truss; these are snow and wind loads. Where the roof is comparatively flat, the **snow load** is estimated at 20 pounds per square foot of roof area; for a roof that has a steep slope, or a rise of more than 12 inches per foot of horizontal distance, it is good practice to assume the snow load to be 12 pounds per square foot of roof area. In the usual style of truss the snow load is conveniently included with the dead load.

4. Wind Pressure.—**Wind pressure** on roofs is always assumed as acting normal (that is, perpendicular) to the slope. In Fig. 1 the outline *a b c* of a roof is shown; the force *d* is normal to the slope *a b*, and represents the assumed pressure of the wind on the roof. The wind generally acts in a horizontal direction, as shown by the full arrow *e*. The maximum horizontal pressure of the wind is always considered to be 40 pounds per square foot; this pressure represents a wind velocity of from 80 to 100 miles per hour, which is a violent hurricane in intensity, and as this velocity is seldom realized and never exceeded except in cyclonic storms, the assumption may be considered reasonably safe. The wind, blowing with a horizontal pressure of 40 pounds, strikes the roof at an angle; consequently, the pressure *d* normal to the slope, is considerably less than 40 pounds.

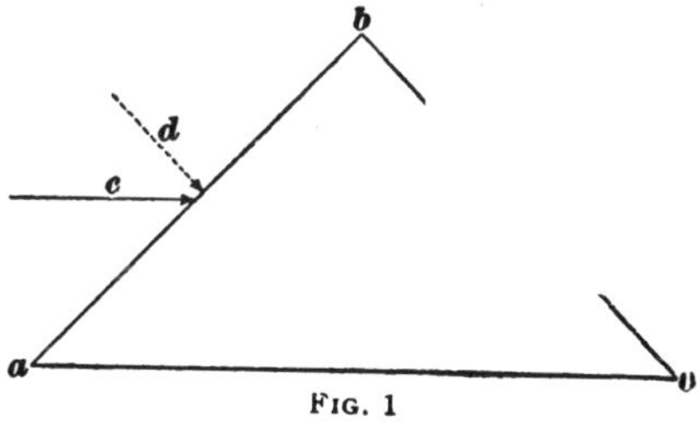

FIG. 1

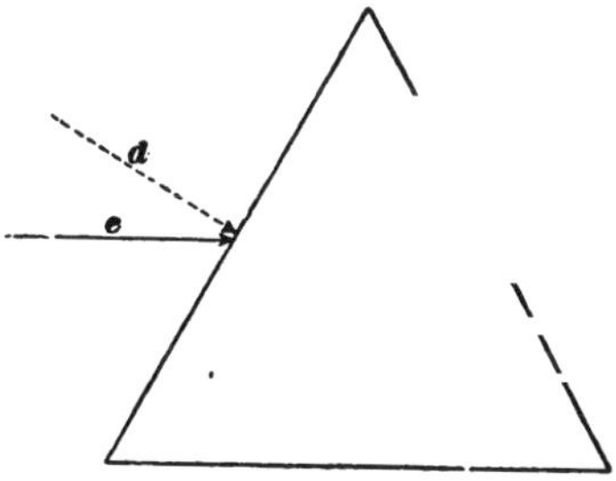

FIG. 2

unless the slope of the roof is very steep. Referring to Figs. 2 and 3 it is clear that the horizontal force *e* of the wind on the slope of the roof, shown in Fig. 2, is almost as intense as on a vertical surface; on the extremely flat roof in Fig. 3, however, the wind exerts hardly any force at all normal to the

FIG. 3

slope, because it strikes the slope at such an acute angle that it has a tendency to slide along and off. The more acute the angle between the lines *e* and *d*, the greater is the pressure normal to the slope; whereas, the greater the distance these lines are apart or the greater the angle, the less is the

TABLE II

WIND PRESSURE NORMAL TO THE SLOPE OF ROOF

Rise Inches per Foot Horizontal	Angle of Slope With Horizontal		Pitch, Proportion of Rise to Span	Wind Pressure Normal to Slope Pounds per Square Foot
	Degrees	Minutes		
4	18	25	$\frac{1}{6}$	16.8
6	26	33	$\frac{1}{4}$	23.7
8	33	42	$\frac{1}{3}$	29.1
12	45	0	$\frac{1}{2}$	36.1
16	53	7	$\frac{2}{3}$	38.7
18	56	20	$\frac{3}{4}$	39.3
24	63	27	1	40.0

pressure normal to the slope, until they form a right angle with each other, where the pressure on the roof may be disregarded. On the basis of a horizontal wind pressure of 40 pounds, the pressure normal to the slope has been reckoned by a formula known as *Hutton's formula*. This formula, being complicated, is not given here, but results derived from it are given in Table II.

5. In order to more fully explain Table II, reference is made to Fig. 4. The rise in the slope *a b* is 6 inches for every 12 inches on the horizontal line *a c*; for instance, at 4 feet from *a* on the horizontal line *a c*, the rise is four times 6 inches, or 2 feet, the angle included between the line of slope *a b* and the horizontal base line *a c* is 26° 33′, and the

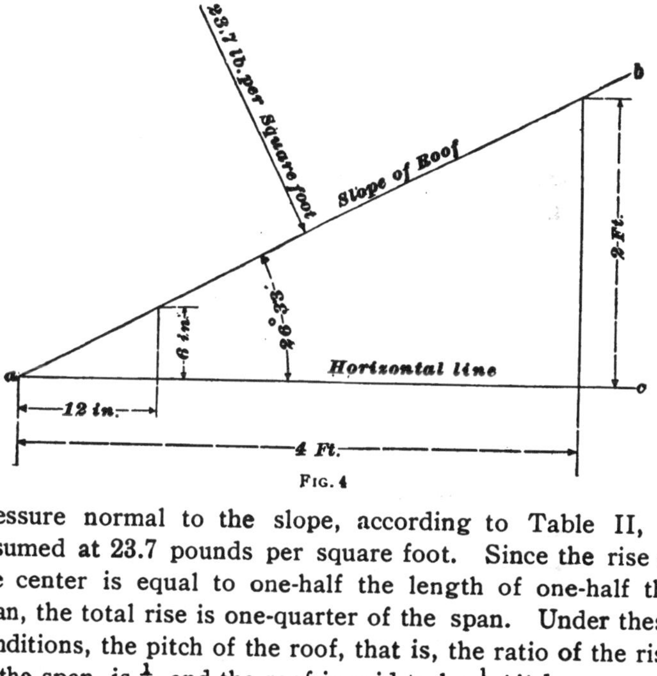

FIG. 4

pressure normal to the slope, according to Table II, is assumed at 23.7 pounds per square foot. Since the rise at the center is equal to one-half the length of one-half the span, the total rise is one-quarter of the span. Under these conditions, the pitch of the roof, that is, the ratio of the rise to the span, is $\frac{1}{4}$, and the roof is said to be $\frac{1}{4}$ *pitch*.

EXAMPLE.—(*a*) What will be the dead load, per square foot of roof surface, on a roof with a 12-inch rise, the span of the trusses being 50 feet, the roof covering 1 inch white-pine sheathing, two layers of Neponset roofing felt, and $\frac{1}{8}$-inch slate 3-inch lap? (*b*) What will be the wind pressure per square foot normal to the slope? (*c*) If the roof trusses are placed 12 feet apart, what will be the entire dead load on one truss? Fig. 5 shows a plan with elevation and detail section of the roof.

SOLUTION.—(*a*) By referring to Table I, it is seen that the approximate weight of a roof truss with a span of 50 ft. is 5 lb. for every square foot of area covered. It is first necessary to obtain the length of the line of slope *a b*; this is done by calculating the hypotenuse of the triangle, or by laying the figure out to scale and measuring. In this case, it is found that *a b* measures about 35 ft. 4 in., equal to 35.33 ft.

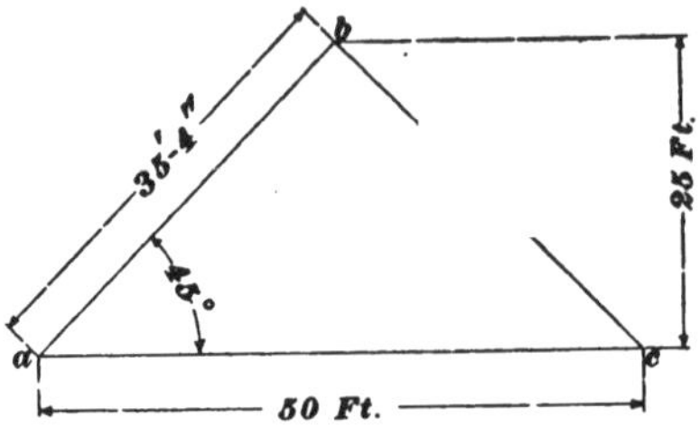

Elevation of Roof

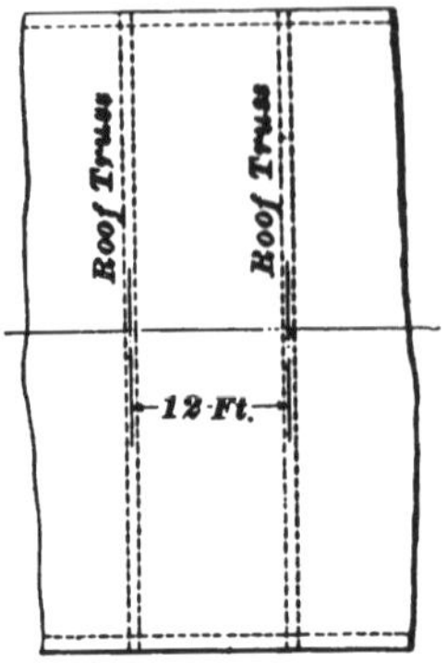

Plan of Roof

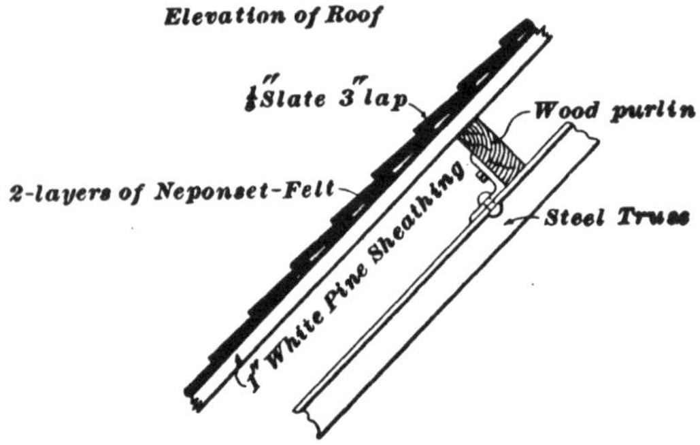

Detail of Roof Covering

FIG. 5

The area covered by the roof supported on one truss is 12 × 50 = 600 sq. ft. The area of the roof supported by one truss is

$$2 \times 35.33 \times 12 = 847.92 \text{ sq. ft.}$$

Then, 600 ÷ 847.92 = .7, which means that the weight of the truss per square foot of roof surface is .7 times 5 lb., or 5 × .7 = 3.5 lb. With the assistance of a table given in *Design of Beams*, the dead load per square foot of roof surface is, then, as follows:

Weight of supporting truss	3.5 lb. per sq. ft.
Weight of white-pine sheathing 1 in. thick . .	2.5 lb. per sq. ft.
Weight of two layers of Neponset roofing paper	.5 lb. per sq. ft.
Weight of slate (⅛ in. thick)	4.5 lb. per sq. ft.
Total	11.0 lb. per sq. ft.

Ans.

The weight of the purlins supporting the sheathing has not been estimated, it being safe in this case to assume that the weight used for the principals, or trusses, is sufficient to cover this item. A snow and accidental load of 12 lb. per sq. ft. of roof surface should also be added to the dead load to get the entire vertical load on the roof.

(*b*) The wind pressure normal to the slope of this roof, according to Table II, for a $\frac{1}{2}$-pitch roof is 36.1 lb., say 36 lb. per sq. ft. Ans.

(*c*) The area of the roof supported by one truss is, as previously found, 847.92 sq. ft., and the dead load is 11 lb. per sq. ft. Then, $847.92 \times 11 = 9{,}327.12$ lb. to be supported by one truss, not including the snow load. Ans.

6. Engineering is not, it must be remembered, an exact science, for the results obtained depend more or less on the judgment and experience of the designer. When, for instance, the wind is blowing a hurricane, snow never lodges on a roof; in such a case, the slates, shingles, and sheathing themselves are exposed and in danger of sudden removal. If, therefore, the full wind pressure is assumed, the snow load may, in most cases, be omitted, especially if it is desired to build an economical roof. It is, however, not well for the beginner to make such assumptions until his experience and judgment is sufficiently developed to enable him to make true deductions.

7. Careful designers sometimes make allowances for the **accidental load** caused by a mass of snow dropping from one roof to another. But this load may usually be ignored, because it is taken care of in the factor of safety, within the limit of which every member in a structure is designed.

EXAMPLES FOR PRACTICE

1. The area of one slope of a $\frac{1}{2}$-pitch roof is 800 square feet; what is the entire pressure on the slope of the roof, provided that the maximum horizontal wind pressure is taken at 40 pounds per square foot?
Ans. 28,800 lb.

2. In a $\frac{1}{4}$-pitch roof, the trusses are 20 feet apart, and the length of the roof slope is 40 feet; what wind load is there on each roof truss, if the horizontal pressure is 40 pounds per square foot?
Ans. 18,960 lb.

3. The purlins supporting a $\frac{3}{4}$-pitch roof are placed 6 feet apart, and the trusses are 12 feet from center to center; what is the maximum load due to the wind on each purlin, provided that the greatest horizontal pressure is 40 pounds per square foot? Ans. 2,830 lb.

GRAPHIC STATICS

8. In the use of graphic statics to determine the stresses on the various members entering into the construction of a frame, not only may the magnitude of the stresses on the members, but the direction in which they act, be determined.

The representation to the eye of the forces existing in the several parts of a frame structure possesses many advantages over their determination by calculation. Graphic analysis being founded on correct principles, the diagrams give results depending for accuracy on the exactness with which the lines have been drawn and on the scale by which they are measured. With ordinary care, the different forces may be obtained much more accurately than the several parts of the frame can be proportioned.

9. Frame and Stress Diagrams.—In Fig. 6, the weight *W* of 1,000 pounds is supported at *c* by the two branching cords *c a* and *c b*, fastened to the pins *a* and *b*. The figure, which is drawn to scale and accurately represents the outline of the structure, is called a **frame diagram.** Now, to obtain the stress on the cords *c a* and *c b*, draw the vertical line *1–2*, Fig. 7; this line must represent the direct pull on the rope, or cord, *c W*, so let each inch represent 400 pounds; then, to represent a force of 1,000 pounds, the length of the line *1–2* shall be 1,000 ÷ 400 = $2\frac{1}{2}$ inches. From *1* draw the line *1–3* parallel to *a c*, and from *2* draw the line *2–3* parallel to the line *c b* of Fig. 6; they will intersect at the point *3* and form a triangle. If the

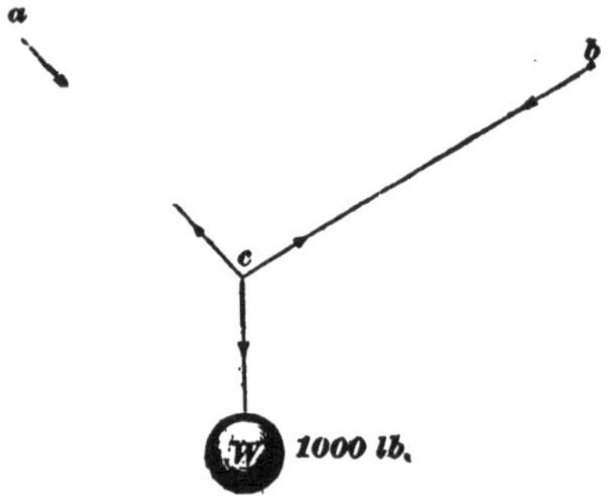

FIG. 6

lines *1–3* and *2–3* are measured with the 1-inch scale, the stresses *ca* and *cb* can be determined.

The diagram, Fig. 7, is called the **stress diagram.** In working out the stresses for a roof truss, it is first necessary to make the frame diagram, drawing it accurately to any scale and representing the outline of the truss and the members of which it is composed. It is then necessary to draw a stress diagram for the dead load on the roof, which sometimes includes the snow load, and another diagram to represent the stresses produced by the action of the wind on the roof.

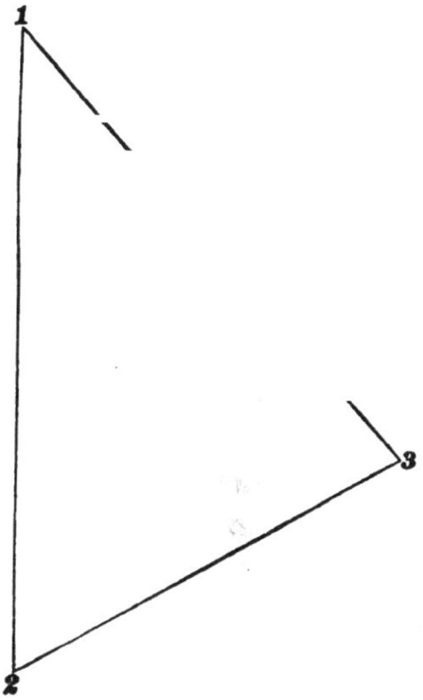

Fig. 7

10. Lettering the Diagrams.—In laying out the frame and stress diagram, it is useful to adopt a system of lettering, so that the relative position of the different members in the frame and stress diagram may be seen at a glance, as in Figs. 8 and 9. In Fig. 8, the reactions at the walls are R_1 and R_2.

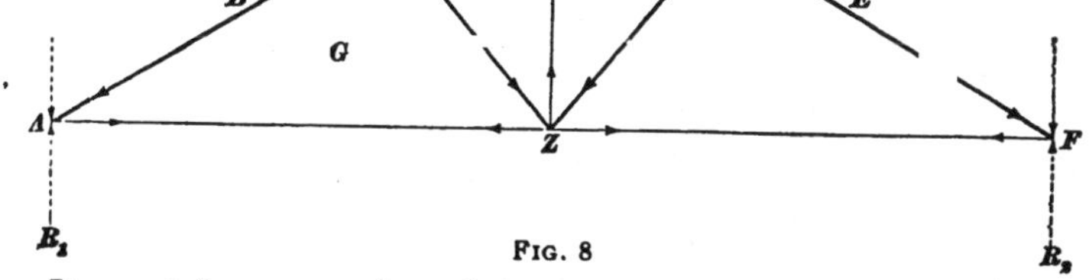

Fig. 8

It must be remembered that a roof truss is nothing more nor less than a beam, and, consequently, the sum of the reactions must be equal to the sum of the loads.

The loads, or forces, acting on a roof truss are always considered as concentrated at the **panel points**, that is, where several members in the structure join one another. In Fig. 8, the loads and reactions on the truss are shown by dotted lines and arrowheads.

The spaces outside of the truss, between the loads, together with each triangle inside of the truss, should be lettered with capitals, as in Fig. 8. It is well to begin at the left-hand reaction of the truss with the letter *A*, working around the outside of the truss in alphabetical order, until the right-hand reaction is reached. Then start with the first triangular space at the left-hand end of the truss, following with the next letter in the alphabet, continuing in alphabetical order, it being well to mark the space between the two reactions, at the center of the truss, *Z*.

It is not absolutely necessary to use this system of lettering, for any letters or numbers may be employed, so long as no two spaces in the truss are marked alike. It is, however, well to have some definite system in engineering work, the foregoing being as convenient as any that can be suggested.

Having marked the truss with letters, as in Fig. 8, the various members and forces may be designated by those letters between which they are located. The left-hand reaction, for instance, will be *Z A*, the first load on the truss *A B*, the second load *B C*, the load at the apex of the truss *C D*, next in succession the loads *D E* and *E F*—always following the truss in the same direction. The last external force on the truss is the right-hand reaction *F Z*.

The lower portion of the left-hand rafter is *B G*, the upper portion *C H*, the left-hand portion of the tie at the base of the truss is *G Z*, the left-hand compression member in the truss is *H G*, and the central tie *I H*, and so on. In designating the various members of the truss, the letters are given in the order in which they occur, always following the joints in the same direction, which in this case is that in which the hands of a watch travel. This being important, it is fully explained further on.

11. The stress diagram for the truss shown in Fig. 8 is represented by Fig. 9. The stress diagram is always drawn to some scale in which the unit measurement represents a certain number of pounds. Thus, suppose 1 inch represents 1,000 pounds, and suppose that in the stress diagram the line *c h* is 4 inches long; then the stress in the member *C H* in Fig. 8 is equal to 4,000 pounds. It will be noticed that small letters are used in the stress diagram and that the letter at the end of a line designates that line. Thus, the length of the line *d i* in the stress diagram, Fig. 9, measures the stress in the corresponding member *D I* of the frame diagram, Fig. 8.

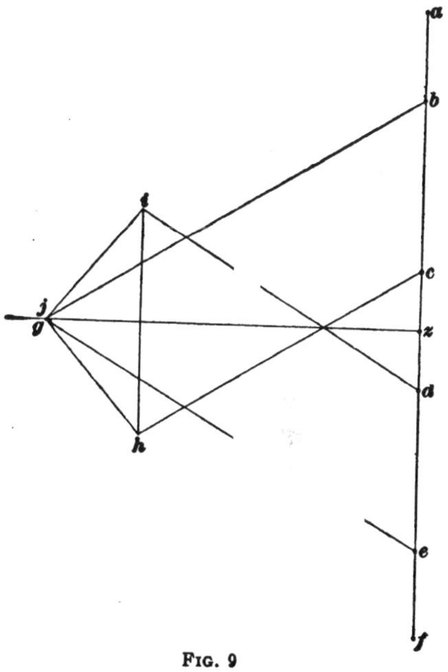

FIG. 9

The distance from *z* to *a* measures the magnitude of the reaction *A Z*; the length of the line *i h* in the stress diagram measures the stress in the corresponding member *I H* of the frame diagram.

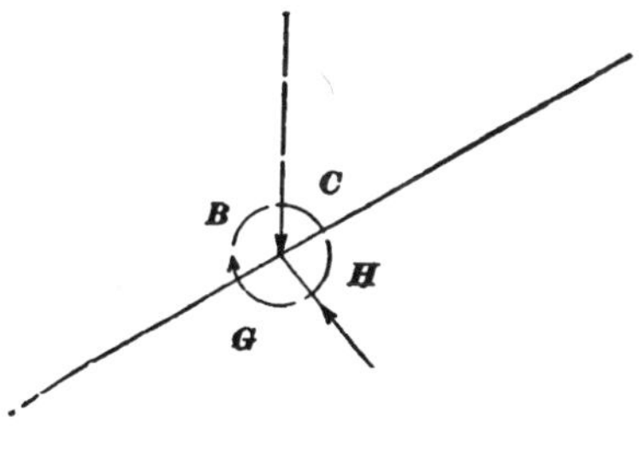

FIG. 10

12. Order of Letters. As previously mentioned, it is very important always to read the letters in the same direction, and in proper order. If, for instance, the joint at the middle of the left-hand rafter (Fig. 10) is examined, the members and stresses must be read off in their proper order, as *B C*,

CH, *HG*, and back again to *GB*. Leaps should not be made from *BC* to *GB*, etc., as this would lead to errors, and prevent the drawing of the stress diagram. Still more important is it to read around the joints in one direction, as in Fig. 10, that is, in the direction of the arrow. By doing this, the following method is devised to find the kind of stress in the members of the frame. Read forces on the stress diagram in the same order as in the frame diagram; thus, *c h* (not *h c*). Then the force acts from *c* to *h* (not from *h* to *c*), which is downwards and to the left. Returning to the frame diagram, it is seen that if the member *CH* exerts a force downwards and to the left on the joint in question it is in compression. Taking the diagonal member at the same joint, since we read *HG* (not *GH*) in the frame diagram, the force in the stress diagram acts from *h* to *g* (not from *g* to *h*), which is upwards and to the left. Returning to the frame diagram, notice that as *HG* pushes, on the joint shown, upwards and to the left it must be in compression. The direction of the other forces acting on the joint under consideration may be found in the same manner. Consider, for instance, the joint at *Z* in the same frame diagram. Here in the diagonal, if the members are read in the same direction as before, it is *GH* (not *HG*). Looking at the stress diagram, it will be observed that *GH* is acting downwards and to the right at the joint *Z*, which shows that the member is in compression. This agrees with the result obtained when examining the other end of this member, for, of course, if there is compression in one part of a member, there must be compression throughout its entire length, since it is only held at the ends.

13. Diagram for Simple Frames.—Fig. 11 shows a force, or load, of 1,000 pounds pulling on the cord *EA*, and held in position by the two cords *AC* and *CE*. Find by graphic statics the stress in these two cords, also the magnitudes of the forces (*AB*, *BC*) (*CD*, *DE*) required to act at the ends of the cords.

Draw the frame diagram, Fig. 11, accurately, say to a scale of $\frac{3}{8}$ inch to 1 foot. Then start to draw the stress diagram, Fig. 12. The force EA in the frame diagram being already known, take some scale, say 400 pounds to 1 inch,

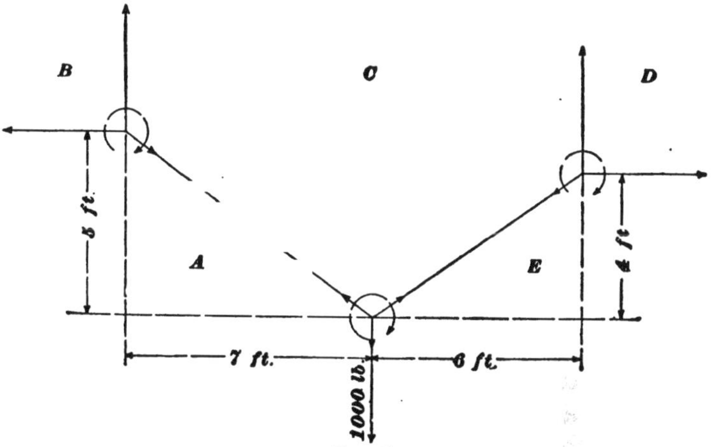

FIG. 11

and draw ea in the stress diagram. It must be drawn parallel to the line along which the force or load EA acts. This force EA being 1,000 pounds, and the scale to which the stress diagram is drawn being 400 pounds to each 1 inch, the line ea must be $2\frac{1}{2}$ inches long. Having drawn ea, work around the joint in the direction of the arrow. From a in the stress diagram draw a line parallel to AC, and from e in the stress diagram draw a line parallel to CE until it intersects the line ac at the point c.

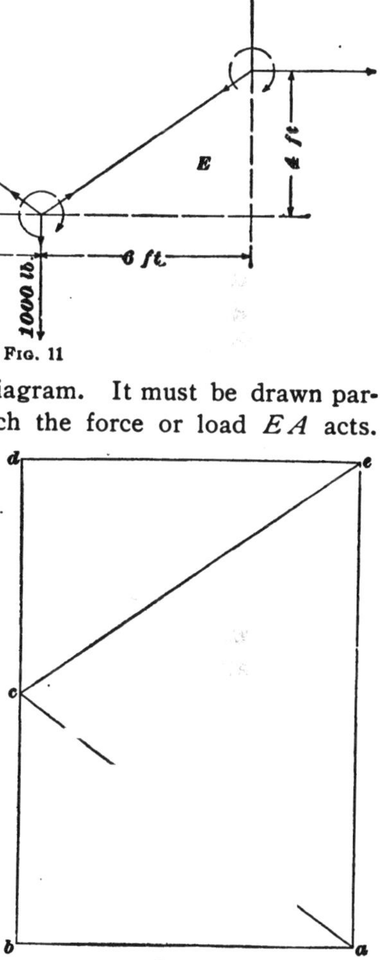

FIG. 12

In going around the stress diagram, the same direction is to be followed. Thus, in going around the joint AE, it is read in the stress diagram from e to a, from a to c, from *c* to *e*, always arriving at the same point from which the start was made. As the stresses are read, their direction should be marked with arrows on the frame diagram, Fig. 11. This shows the direction of the stress and designates whether it is compression or tension. Forces acting away from a joint are always tensile stresses; those acting toward a joint are always compressive stresses.

If there is tension at one end of a member, it is evident there must be an equal amount of tension at the other end; if there is compression at one end of a member, there is an equal amount at the other.

To continue the solution of the problem in Fig. 11: Having gone around the joint *E A C*, proceed to go around the joint *A B C* in the same direction. Having the point *a*, draw from it a line parallel to *A B* in the frame diagram, then draw a line from *c* parallel to the line *B C* in the frame diagram; the point where the two lines intersect is *b*. The polygon of forces may be read from *c* to *a*, from *a* to *b*, from *b* to *c*, always moving in the same direction and arriving at the starting point. The next joint to work around is the joint *C D E*. Starting at the point *c* already determined, draw a line parallel to the direction of *C D*, and from *e* draw a line parallel to *D E*; the polygon of forces is from *a* to *c*, *c* to *d*, and *d* to *e*, the last point in the diagram, and the point from which the whole stress diagram was started. Then by measuring the lines (with the 1-inch scale, wherein every inch represents 400 pounds) in the stress diagram, the stresses in the different members of the frame diagram may be found. If, for instance, it is desired to obtain the stress in the cords *A C* and *C E*, measure the lines *a c* and *c e* in the stress diagram; likewise, to obtain the forces *A B*, *B C*, *C D*, and *D E*, measure the length of the lines *a b*, *b c*, *c d*, and *d e* in the stress diagram.

14. In Fig. 13 is shown a case similar to that in the preceding article. The compression member *B D* and the tension members *A D* and *D C* form a triangular frame, which supports the downward pull of 1,000 pounds. The triangular frame is supported, in turn, by the reactions *AB* and *BC*. Draw the stress diagram to determine the stress in the various members.

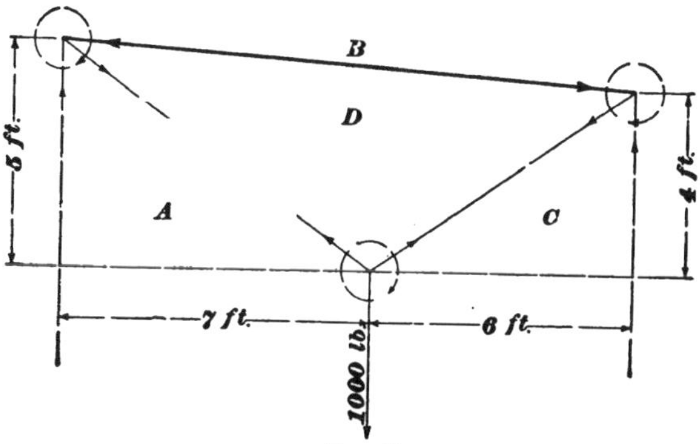

FIG. 13

Take a scale, in this case 100 pounds to $\frac{1}{4}$ inch, and draw the vertical line *ca*, Fig. 14, equal to 1,000 pounds. This line represents the force *CA* in the frame diagram, Fig. 13. Start to work around the joint *ADC*, in the direction of the arrow. The first member encountered is *AD*. Hence, from *a* in the stress diagram draw a line parallel to *AD* in the frame diagram. Then, *DC* being the next member met with, from *c* in the stress diagram draw a line parallel to *DC*. The point of intersection of the two lines just drawn is *d*. This done, go around the joint again, to see that none of the members have been omitted, and also to get the direction in which the stresses act. Starting at *c* in the stress diagram, and going around the joint *CAD*, the polygon of forces is as follows: from *c* to *a*, from *a* to *d*, and from *d* back

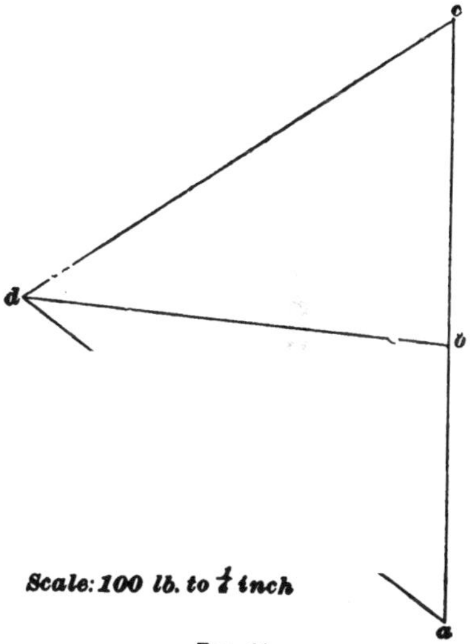

FIG. 14

again to c, thus arriving at the point from which the start was made. The next joint in the frame diagram is ABD. The point b on the line ca is not known, but may be determined by calculating the reactions AB and BC in the same manner as for a beam. Thus, the load of 1,000 pounds is placed on the assumed beam, 6 feet from the reaction BC. The moment about CB is $1{,}000 \times 6 = 6{,}000$ foot-pounds, while the reaction at AB equals $6{,}000 \div 13 = 461$ pounds. Knowing that the force AB is 461 pounds and that it acts upwards, the point b can easily be located by measuring from a on the line ac in the stress diagram; then the line bd may be drawn

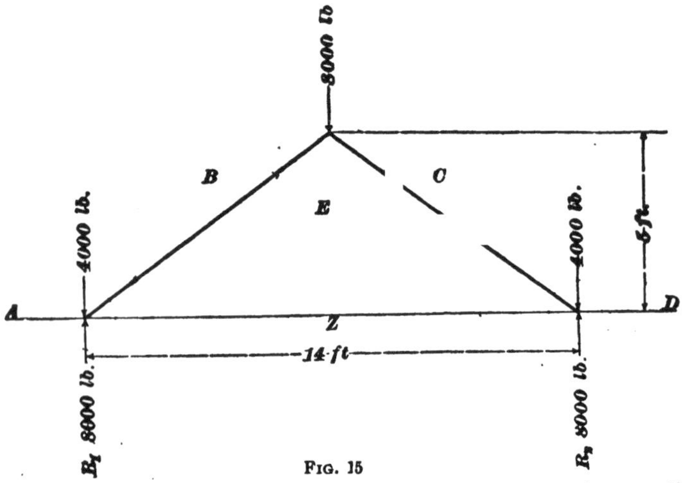

FIG. 15

and if found parallel to the member BD in the frame diagram, the stress diagram is correct. In this case, however, it is not necessary to calculate the reactions AB and BC, the point d having been already determined, and since we know that the line db must be parallel to DB, all that is needed to locate the point b is to draw a line from d parallel to DB, and the point where it cuts the line ca is b. Having found the point b and drawn the line db, go around the joints ABD and CDB, marking the direction of the stress by the arrowheads, as shown in the frame diagram, Fig. 13. Around the joint ABD the polygon of forces is from a to b, from b to d, and from d back again to a. Working around the joint *C D B*, the polygon of forces is from *c* to *d*, from *d* to *b*, and from *b* back again to *c*. This completes the stress diagram; the magnitude of the stresses in the several members of the frame diagram is found by measuring the corresponding lines in the stress diagram.

15. Diagram for a Small Roof Truss.—Fig. 15 is the frame diagram for a small roof truss. Each of the rafter members *E B* and *C E* is connected at its foot by the tension member *E Z*. The loads and their reactions are as shown in the frame diagram. Determine the stresses in the several members composing the truss.

Draw the vertical line *a d*, shown in the stress diagram, Fig. 16. Lay off to any scale, say, in this case 2,000 pounds to ½ inch, the load *a b*; then, to the same scale, the loads *b c* and *c d*. From the point *d*, the reaction *d z*, 8,000 pounds, acts upwards, which determines the point *z*. Now go around the joint *A B E Z*. The reaction *Z A* acts upwards and *A B* downwards. Then, from the point *b* in the stress diagram draw the line *b e* parallel to *B E* in the frame diagram, and from *z* draw the line *e z* parallel to the member *E Z* in the frame diagram. The point of intersection will be the point *e*. Having gone thus far, again go around the joint, to get the direction of the stress in the members and to see whether the polygon of forces is correctly drawn. Go, for instance, from *z* to *a* upwards; *a* to *b* downwards; then from *b* to *e*, and from *e* back again to *z*, the starting point. The next

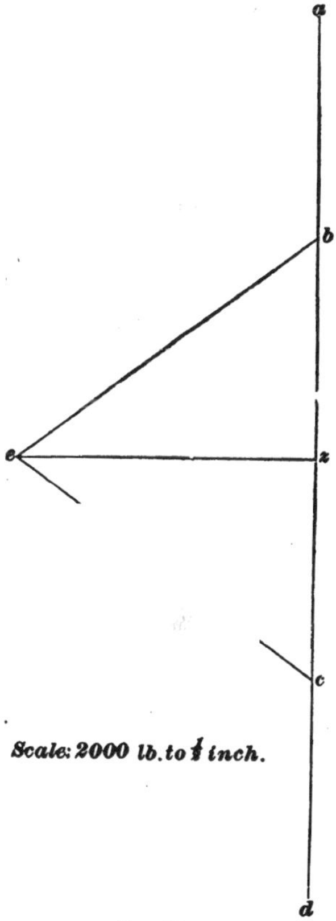

FIG. 16

joint in the frame diagram is BCE. The force bc being already determined, from point c, draw the line ce, parallel to the member CE in the frame diagram; this line passes through the point e if the diagram has been drawn correctly. The polygon of forces at EBC is from b to c downwards, then from c to e, and back again from e to b, the starting point. The next joint in the frame diagram is $ECDZ$, and the polygon of forces in the stress diagram is from e to c already drawn, from c to d, d to z, and then from z back to e.

The stress diagram completed, all that remains is to measure the various lines in the stress diagram that represent the corresponding members in the frame diagram. Thus, eb measures $1\frac{3}{4}$ inches, the scale being 2,000 pounds to $\frac{1}{8}$ inch; hence, the stress in this member is 7,000 pounds; the line ez measures about $1\frac{3}{8}$ inches, and the stress in the member ez is 5,500 pounds. In this manner, the stress in any member may be determined.

16. Diagram for a Jib Crane.—A jib crane proportioned as in Fig. 17 has a load of 30,000 pounds suspended

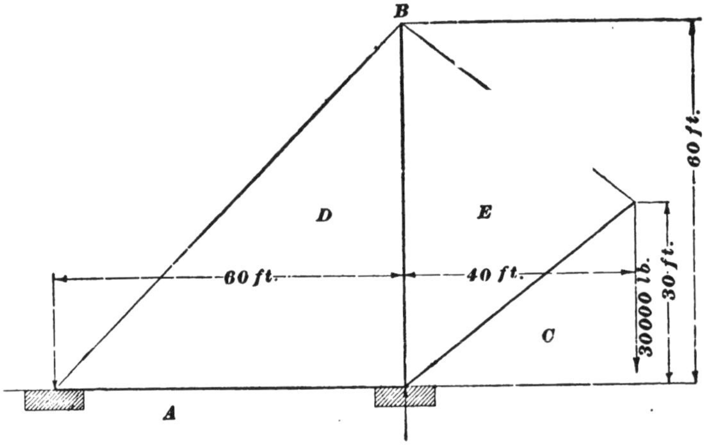

FIG. 17

at the end of the jib; what are the stresses in the guy ropes and in the different members of the crane, and what are the reactions CA and DA?

In the stress diagram, Fig. 18, draw the vertical line *bc* equal to 30,000 pounds, and from the point *c* draw the line *ce* parallel to *CE* in the frame diagram, Fig. 17. Then from *b* draw the line *eb* parallel to *EB* in the frame diagram. Again going around the joint to check the polygon of forces, they are found to be from *c* to *e*, from *e* to *b*, and from *b* back again to *c*. The next joint encountered is *EDB*. Hence, from *e* draw *ed* upwards parallel to *ED*, and from *b* draw *db* parallel to *DB*, the point where these two lines intersect being *d*. The polygon of forces about the joint *EDB* is from *b* to *e*, and from *e* to *d*, and from *d* back again to *b*, the starting point. Next, go around the joint *CADE*, and draw *ca* upwards; then from *d* draw *ad* parallel to *AD* in the frame diagram; where the lines just drawn intersect will be the point *a*; *de* and *ec* have already been drawn. The remaining joint to work around is *ABD*.

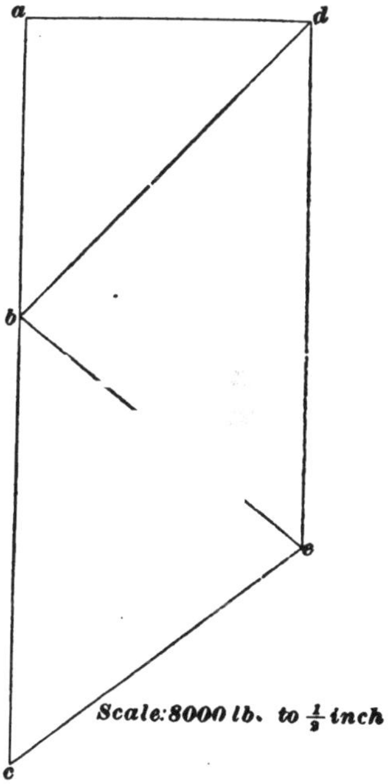

FIG. 18

On looking at the stress diagram, it may be seen that the forces around this joint have already been determined, completing the stress diagram.

The stresses in the members may be determined, as already stated, by measuring the lines corresponding to them in the stress diagram, with the scale to which the diagram has been drawn.

17. Roof Truss With a 40-Foot Span.—Fig. 19 shows the frame diagram for a 40-foot span roof truss The loads are as shown, the compression members being indicated by heavy lines, and the tension members by light lines. Required, to draw the stress diagram for this truss.

First, draw the vertical line *af* as shown in the stress diagram, Fig. 20; mark the point *a* and lay off on this vertical line, to any scale, using, for instance, 4,000 pounds to ½ inch, the loads *ab*, *bc*, *cd*, *de*, and *ef* corresponding to the loads *AB*, *BC*, *CD*, *DE*, and *EF* in the frame diagram. The truss being symmetrically loaded, the loads are the same in amount on the two sides of the center line. The reactions R_1 and R_2 are therefore each equal to one-half of the load, in this case, 16,500 pounds. Hence, *za* may

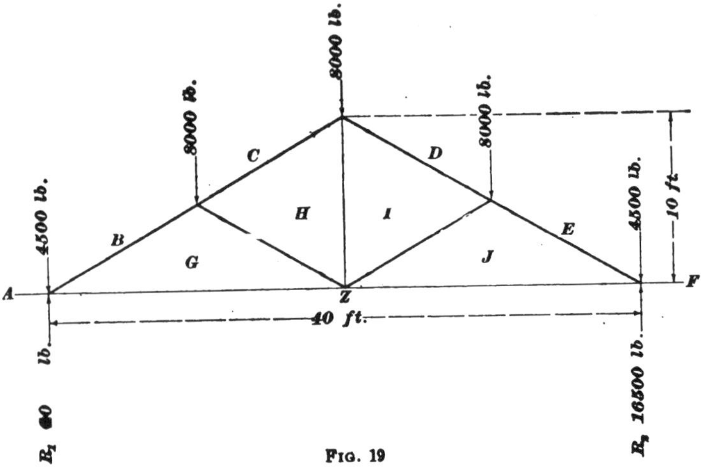

FIG. 19

be laid off on the vertical line, and as R_1 equals R_2, *za* must equal *fz*; consequently, *z* is located centrally between *a* and *f*, or between *c* and *d*. The point *z* having been determined, proceed with the diagram by going around the joint *ABGZ*. Draw *bg* in the stress diagram parallel to *BG* in the frame diagram; then from *z* draw *gz* parallel to *GZ*, the point where the two lines intersect being *g*. The next joint is *BCHG*. As *bc* in the stress diagram is already known, draw *ch* parallel to *CH* and *hg* parallel to *HG*. Then the polygon of forces around this joint will be from *b* to *c*, from *c* to *h*, from *h* to *g*, and from *g* back again to *b*. It is now expedient to analyze the joint *CDIH*. In the stress diagram, *hc* and cd have already been obtained; then from d, draw di parallel to DI, and from the point h, already known, draw ih parallel to IH. The polygon of forces around this joint will be from c to d, from d to i, from i to h, and h to c, the starting point, the direction marking the direction in

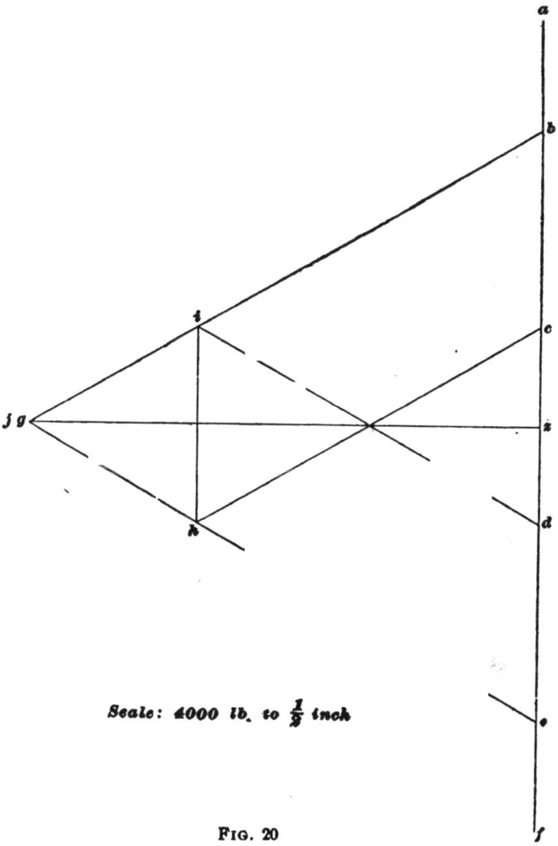

FIG. 20

which the stresses act in the stress diagram, on the members in the frame diagram. Now analyze the forces around the joint $GHIJZ$ in which the forces zg, gh, and hi have been obtained. From the point i, ij is drawn parallel to IJ; and from z, jz is drawn parallel to JZ; the point j is found to fall on the point g, and the polygon of forces around this joint is from *z* to *g*, from *g* to *h*, from *h* to *i*, from *i* to *j*, and from *j* back again to *z*, the starting point. The stress in the members around the joint *IDEJ* should next be determined, *ji*, *id*, and *de* being already known. From *e* draw *ej* parallel to *EJ*. The polygon of forces around this joint will then be from *i* to *d*, from *d* to *e*, from *e* to *j*, and from *j* back again to *i*.

The only remaining joint to go around is *JEFZ*. By referring to the stress diagram, it is seen that the stresses in these members have been determined, while the polygon of forces around this joint is from *j* to *e*, from *e* to *f*, from *f* to *z*, and back again from *z* to *j*.

The stress diagram completed, the magnitudes of the stresses may be determined by measuring the various lines with the scale to which the diagram has been drawn, in this case 4,000 pounds to ½ inch.

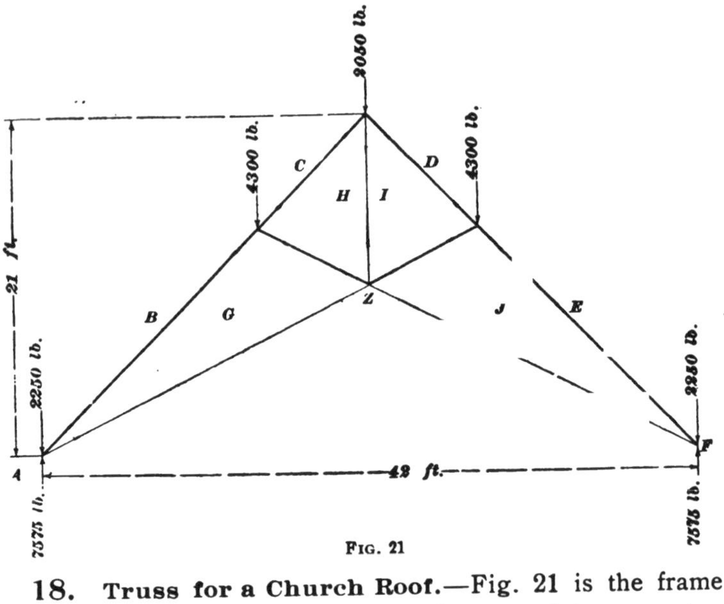

FIG. 21

18. Truss for a Church Roof.—Fig. 21 is the frame diagram of a form of truss sometimes used to support a church roof. Determine the stress diagram for the dead load and also the stress diagram for the wind pressure on this roof.

First, draw the stress diagram (Fig. 22) for the dead load. As the dead loads on the truss are symmetrical both in amount and location with regard to the center line of the truss, the reactions are the same at either end of the truss, and each one is equal in amount to one-half of the load, in this case 7,575 pounds.

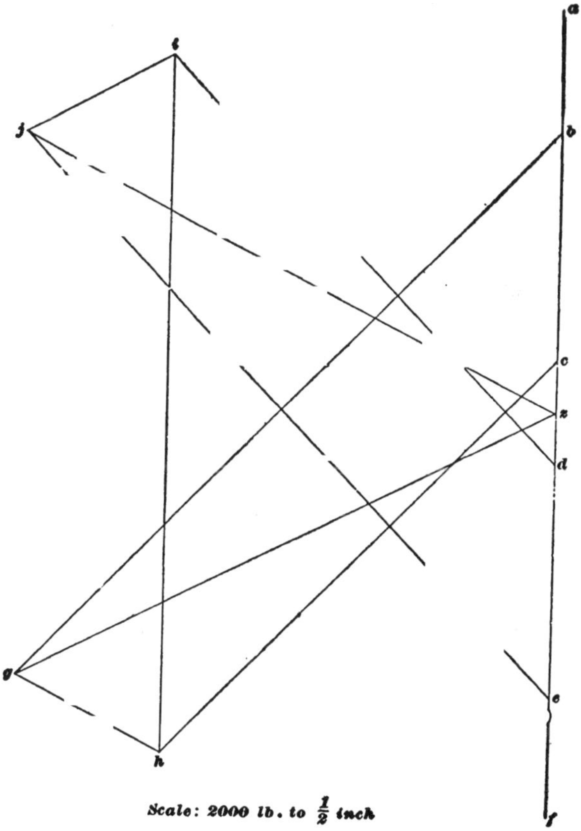

Scale: 2000 lb. to $\frac{1}{2}$ inch

FIG. 22

Draw the vertical line *a f* in the stress diagram; then, starting at the point *a*, lay off the scale of, say, 2,000 pounds to every $\frac{1}{2}$ inch, the force *a b* equal to *A B* in the frame diagram; then lay off *b c* equal to *B C*, *c d* equal to *C D*, *d e* equal to *D E*, and *e f* equal to *E F*. Then, as the truss is symmetrically loaded, the point *z* is located midway between the points *a* and *f*. If the truss were not symmetrically loaded, the reactions would have to be calculated in the same manner as in a beam.

Having located the loads and their reactions on the vertical line *af*, obtain the stresses in the members around the joint *ABGZ* from the point *b*, by drawing a line *bg* parallel to *BG* in the stress diagram; then from *z* draw the line *gz* parallel to *GZ*, and the intersection of these two lines will be the point *g*. The polygon of forces around this joint is from *b* to *g*, from *g* to *z*, from *z* to *a*, and then from *a* back to *b*, the starting point. Bear in mind that the forces in the stress diagram representing the reactions must have the same direction as the reactions in the frame diagram. The lines determining the stresses around the joint *BCHG* should next be drawn; *bc* having been determined, from *c* draw a line *ch* parallel to *CH*, and from *g* draw a line *gh* parallel to *HG*, the intersection of these two lines determining the point *h*; the polygon of forces is from *b* to *c*, from *c* to *h*, from *h* to *g*, and back again from *g* to *b*.

Now work around the joint *CDIH*; *cd* being already known, from the point *d* draw the line *di* parallel to *DI*, and from the point *h* draw the line *hi* parallel to *IH*, the intersection of the two lines being the point *i*. In going around the joint *GHIJZ*, the stresses in the members *zg* *gh*, and *hi* have already been determined and drawn in the stress diagram. Then from the point *i* draw the line *ij* parallel to *IJ*, and from *z* draw the line *zj* parallel to *JZ*, and the intersection of these two lines will be the point *j*. The polygon of forces around this joint is from *z* to *g*, from *g* to *h*, from *h* to *i*, from *i* to *j*, and from *j* back to *z*, the starting point.

The next joint to analyze, in going around the truss, is *DEJI*; *ji*, *id*, and *de* being known, the only remaining force to determine is the stress in the member *EJ*. The point *e* being fixed, draw the line *ej* parallel to *EJ* in the frame diagram, and if this line, which completes the diagram, passes through the point *j*, the diagram is correct and accurately drawn. The stresses around the right-hand heel of the truss are all known, the line *ej* just drawn having been the only unknown member at this joint.

The polygon of forces around the joint *DEJI* is from *d* to *e*, from *e* to *j*, from *j* to *i*, and from *i* to *d*, the starting point. The polygon of forces around the joint *EFZJ* is from *e* to *f*, from *f* to *z*, from *z* to *j*, and from *j* back again to *e*, completing the stress diagram for the dead, or vertical, load on the roof truss.

19. The wind diagram, however, remains to be drawn. Redraw the frame diagram as shown in Fig. 23. The wind is always considered as acting normally, or at right angles, to the roof, the amount of its pressure at the different joints

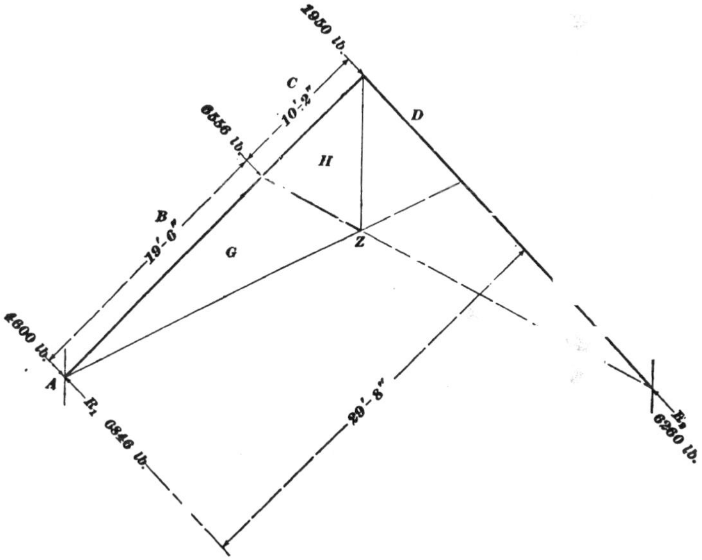

FIG. 23

of the truss being shown on the frame diagram. As the heels of the trusses are fixed, the reactions act in lines parallel to the wind pressure.

To estimate the magnitude of the reactions R_1 and R_2, consider the left-hand rafter member as a beam, and R_1

and R_2 as the reactions supporting it. The moments due to the wind pressure BC and CD acting about R_1 are:

At BC,	$6{,}556 \times 19.5$	=	127,842 foot-pounds
At CD,	$1{,}950 \times 29.66$	=	57,837 foot-pounds
Total,			185,679 foot-pounds

The lever arm with which R_2 resists the wind pressure acting at the joints, since the apex of the truss happens in

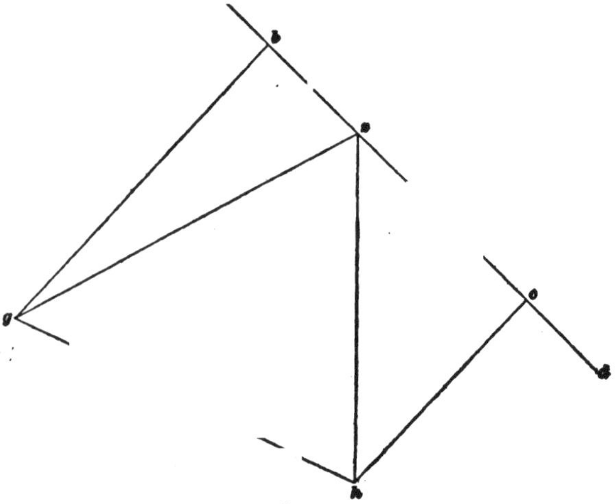

Scale: 2000 lb to $\frac{1}{8}$ inch

FIG. 24

this case to be a right angle, is 19 feet 6 inches + 10 feet 2 inches = 29 feet 8 inches, so $185{,}679 \div 29.66 = 6{,}260$ pounds, the amount of the reaction at R_2. The sum of the loads being

$$4{,}600 + 6{,}556 + 1{,}950 = 13{,}106 \text{ pounds,}$$

the reaction at R_1 is $13{,}106 - 6{,}260 = 6{,}846$ pounds.

First, draw the load line ad in the stress diagram, Fig. 24; this line is parallel to the direction of the wind pressure and the reactions. Now lay off to the scale to which the stress diagram is drawn—in this case 2,000 pounds to every ½ inch—the force *a b* equal to *A B* in the frame diagram; then lay off *b c* equal to *B C* and *c d* equal to *C D*. From *d* lay off the magnitude of the reaction R_2, or *d z*, which determines the point *z*, and the distance *z a*, according to scale, represents the left-hand reaction R_1. The polygon of external forces is, then, from *a* to *b*, from *b* to *c*, from *c* to *d*, from *d* to *z*, and from *z* back again to *a*, the starting point. This polygon, as may be readily seen, is a straight line, as in all cases so far analyzed.

Continue the stress diagram, Fig. 24, by going around the joint *A B G Z*; from the point *b* draw the line *b g*, and from the point *z* draw the line *z g* parallel to the corresponding members in the frame diagram, the point where the two lines intersect being *g*. The polygon of forces around this point is from *a* to *b*, from *b* to *g*, from *g* to *z*, and from *z* back again to *a*.

The next joint is *B C H G*; *b c* has been already obtained; then from the point *c* draw the line *c h* parallel to *C H*, and from *g* draw the line *g h* parallel to *H G*, *h* being the point where these two lines cross. Disregard the two members in the frame diagram shown in dotted lines, which do nothing toward sustaining the wind pressure. Now work around the joint *H C D Z*; *c d* and *d z* are known; draw from *z* the line *z h*, parallel with *Z H* in the frame diagram; if this closing line of the diagram passes through the point *h*, the diagram has been drawn accurately. The polygon of forces around the joint *B C H G* is from *b* to *c*, from *c* to *h*, from *h* to *g*, and from *g* back again to *b*. The polygon of forces around the joint *C D Z H* is from *c* to *d*, from *d* to *z*, from *z* to *h*, and from *h* back again to *c*. The polygon of forces around the joint *Z G H* is from *z* to *g*, from *g* to *h*, and from *h* to *z*, the starting point.

The stress diagram for both the dead and the wind load being complete, to obtain the stress in each member of the truss, it is required to determine, by scale, the stress due to both the dead and wind loads in each member, adding the two together for the maximum load in the member. To determine, for instance, the stress in the strut *H G*, measure the length of the line *hg* in the stress diagram, Fig. 22, for the dead load; then measure the same line *hg* in the stress diagram, Fig. 24, for the wind load, add the two measurements together, and determine the maximum stress in the strut *hg* from the assumed scale of the drawing.

It must be remembered that while the wind acting on one side of a truss does not create stresses in all the members on the opposite side, these members should be proportioned in like manner as the other members, because the wind is quite as likely to blow on this side of the roof and reverse the conditions.

20. Howe Truss With an 80-Foot Span.—Fig. 25 is the frame diagram for an 80-foot span Howe roof truss It is desired to draw the dead-load diagram and the wind-stress diagram.

Draw the frame diagram shown in Fig. 25 and mark the dead load coming on the different panel points, or joints, in the truss. The truss being symmetrically loaded, the reactions R_1 and R_2 are each equal to one-half the load on the truss.

Draw the stress diagram, Fig. 26, for the dead load, say to the scale of 4,000 pounds to ½ inch. Draw the vertical load line *a j*, and determine the point *z*, having previously located on the line all the loads. Then draw the stress diagram by the methods previously given. Only one-half of the diagram need be drawn, as the stresses obtained on one side of the center of the truss apply to the other side. For instance, *f q* is the same as *p e*. Having completed one-half of the stress diagram for the dead load, the frame diagram should be redrawn as shown in Fig. 27. The direction and amount of the wind pressure at the several panel points, or joints, of the truss, are shown in the frame diagram.

As both ends of the truss are secured against sliding, the reactions act in a direction parallel to the wind pressure. If the left-hand side of the truss is secured, the right-hand side

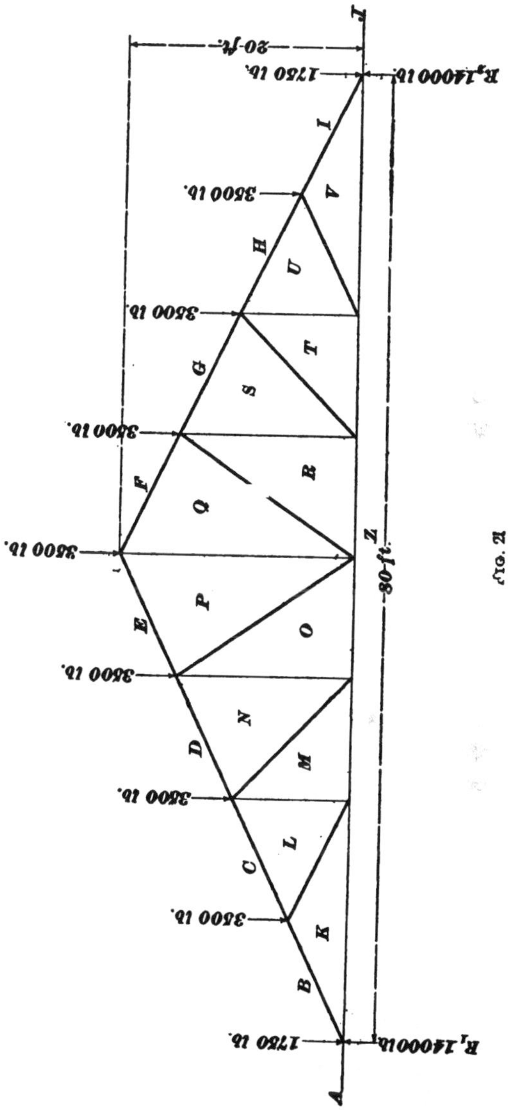
1750 lb.
3500 lb.
3500 lb.
3500 lb.
3500 lb.
3500 lb.
3500 lb.
3500 lb.
1750 lb.
R_1 14000 lb.
R_2 14000 lb.
20 ft.
80 ft.
A B C D E F G H I J
K L M N O P Q R S T U V
Z

FIG. 21

being on rollers, as is sometimes the case with iron or structural-steel trusses, to allow for expansion, then the right-hand reaction, instead of being parallel to the direction of the wind, will be vertical. This makes considerable difference in the stress diagram, as will be explained further on.

To determine the magnitude of the reactions R_1 and R_2, let R_1, Fig. 27, be the center around which the moment

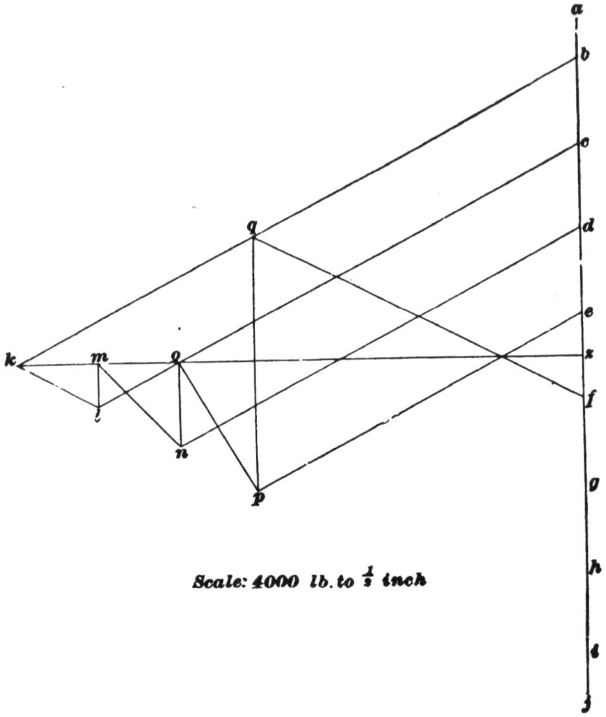

FIG. 26

of R_2 is taken; then the perpendicular distance between the line of action of R_2 and the point R_1 will be 71.22 feet. Extend the left-hand rafter until it cuts the line of action of the force R_2 at the point y'. Regard this extension and the rafter as a beam, and calculate the magnitude of the reactions R_1 and R_2 by the methods given for beams.

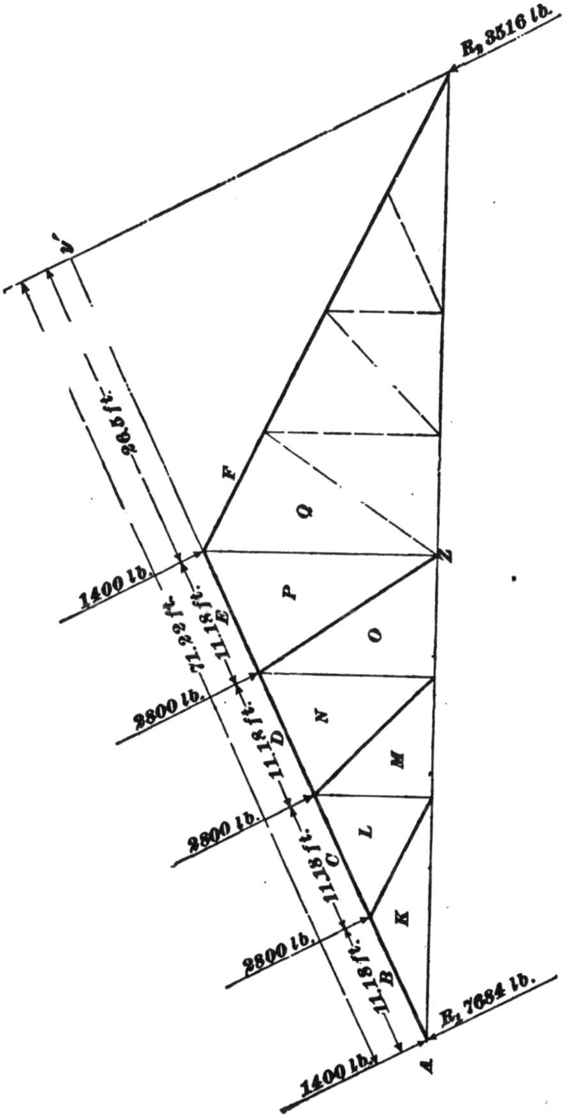
R2 3516 lb.
R1 7684 lb.
1400 lb.
2800 lb.
2800 lb.
2800 lb.
1400 lb.
26.5 ft.
71.22 ft.
11.18 ft.
11.18 ft.
11.18 ft.
11.18 ft.
A
B
C
D
E
F
K
L
M
N
O
P
Q
Z

FIG. 27

The moments about R_1 are as follows:

$$2{,}800 \times 11.18 = 31304 \text{ foot-pounds}$$
$$2{,}800 \times 22.36 = 62608 \text{ foot-pounds}$$
$$2{,}800 \times 33.54 = 93912 \text{ foot-pounds}$$
$$1{,}400 \times 44.72 = 62608 \text{ foot-pounds}$$
$$\text{Total,} \quad 250432 \text{ foot-pounds}$$

and $250{,}432 \div 71.22 = 3{,}516$ pounds, the reaction R_2. Having found R_2, find R_1 by subtracting R_2 from the sum of the loads. The sum of the loads is

$$1{,}400 + 2{,}800 + 2{,}800 + 2{,}800 + 1{,}400 = 11{,}200 \text{ pounds}$$

Then, $11{,}200 - 3{,}516 = 7{,}684$ pounds, the reaction R_1.

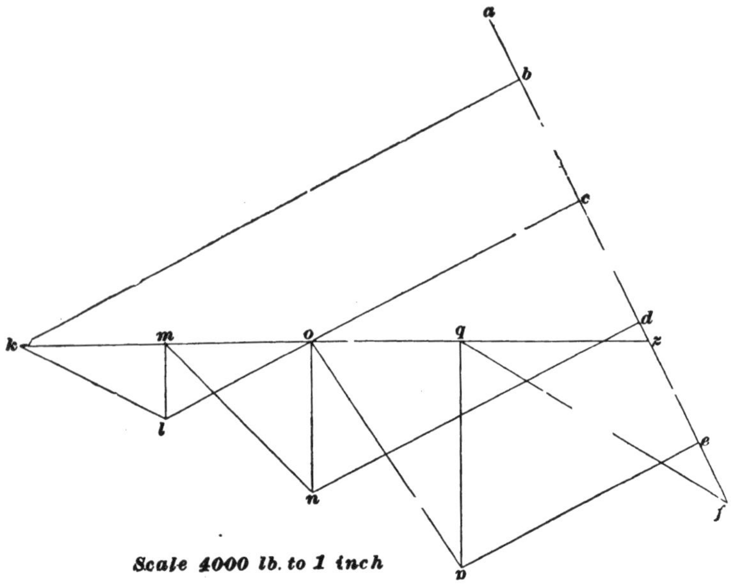

Fig. 28

Next, lay out the wind diagram, Fig. 28, by drawing the road line *a f* parallel to the direction of the wind in the frame diagram, Fig. 27. Lay off to the scale (in this case 4,000 pounds to 1 inch) the forces *a b*, *b c*, *c d*, *d e*, and *e f* equal to *AB*, *BC*, *CD*, and so on, in the frame diagram. Then, from *a* lay off *a z* equal to the reaction *ZA*, or R_1. If the other forces or loads have been laid off accurately, *fz* should, on measurement, be found equal to the right-hand reaction R_2.

The first joint to analyze is *ABKZ*. Start at *b* and draw the line *bk* parallel to *BK* in the frame diagram; then from *z* draw *zk* parallel to *KZ*, where the two lines intersect being the point *k*. Then the polygon of forces around this joint is from *a* to *b*, from *b* to *k*, from *k* to *z*, and from *z* back again to the starting point *a*.

The next joint to analyze is *BCLK*. From the point *c* draw the line *cl* parallel to *CL* in the frame diagram, and from *k* draw the line *kl* parallel to the member *LK*, the point of intersection being *l*. The polygon of forces around this joint is from *b* to *c*, from *c* to *l*, from *l* to *k*, and from *k* back again to *b*.

To analyze the joint *KLMZ*: *kl* being already known, the next number is *LM*; therefore, from the point *l* draw the line *lm* parallel to *LM* in the frame diagram. As the next member around this joint is *MZ*, to which *mz* in the stress diagram is parallel, the point *m* is located where the line *lm* intersects the line *mz*; this completes this joint, the polygon of forces around it being from *k* to *l*, from *l* to *m*, from *m* to *z*, and from *z* back again to *k*.

To determine the stresses in the members around the joint *CDNML*, draw from the point *d* the line *dn*, parallel to the member *DN* in the frame diagram; then, from the point *m* draw *mn* parallel to *NM*. The polygon of forces around this joint is from *c* to *d*, from *d* to *n*, from *n* to *m*, from *m* to *l*, and from *l* back again to *c*.

To analyze the forces around the joint *MNOZ*, draw from *n* the line *no* upwards, parallel to *NO* in the frame diagram; as the next member *OZ* is horizontal, the point *o* must be at the intersection *no* and *oz*. This completes this joint, and the polygon of forces around it is from *m* to *n*, from *n* to *o*, from *o* to *z*, and from *z* back again to *m*, the starting point.

Now analyze the joint *DEPON*. From *e* draw the line *ep* parallel to *EP* in the frame diagram, and from *o* draw the line *op* parallel to *PO*. The intersection of these two lines determines the point *p*, and the polygon of forces around this joint is from *d* to *e*, from *e* to *p*, from *p* to *o*, from *o* to *n*, and from *n* back again to *d*, the starting point.

The analysis of the joint *OPQZ* is made by drawing the vertical line *pq* from the point *p*; the point where *pq* intersects *qz* is *q*. Then the polygon of forces around this joint is from *o* to *p*, from *p* to *q*, from *q* to *z*, and from *z* back again to *o*. The members shown in dotted lines do not sustain any stresses from the pressure of the wind, when it blows on the left-hand side of the truss.

The final joint to consider, thus completing the stress diagram, is *EFQP*. There is only one unknown force around this joint, and that is the stress in the member *FQ*. A line drawn from *f* in the stress diagram, parallel with the member *FQ*, should pass through the point *q*; if it does not, the diagram has been inaccurately drawn. This is always a test of the accuracy of the stress diagram, and if the last line in this diagram does not close on the proper point, when drawn parallel to the member it represents, there is something so radically wrong as to demand that the stress diagram be redrawn, to determine whether the loads and reactions have been laid out correctly, and whether any of the joints or members in the structure have been passed over.

It can be readily seen that there are no stresses produced in the internal members of the truss on the side opposite to that from which the wind is blowing. First, examine the joint at the right-hand reaction; the stresses in *FQ* and *QZ* can be found. Next, examine the first joint on the rafter above the right-hand reaction; it will be seen that one internal member terminates at this joint. This member can have no stress in it because the other two members of the joint are in the same line and there is no fourth member to oppose that part of the stress of the internal member that is at right angles to the rafter and thus keep the joint in equilibrium. As there is no stress in the first internal member, by like reasoning about the first joint of the lower chord, it can be proved that there is no stress in the second internal member, nor in the third, and so on to the member *PQ*.

21. The two diagrams completed, measure the different lines and obtain the stresses in the various members. tabulating the results as follows:

Member	Dead-Load Diagram Pounds	Wind-Load Diagram Pounds	Total of Both Pounds	Kind of Stress
BK	27,000	12,000	39,000	Compressive
CL	23,500	10,000	33,500	Compressive
DN	19,500	7,600	27,100	Compressive
EP	16,000	5,680	21,680	Compressive
KZ	24,000	13,300	37,300	Tensile
MZ	21,000	10,000	31,000	Tensile
OZ	17,500	7,000	24,500	Tensile
LK	4,000	3,500	7,500	Compressive
NM	5,000	4,400	9,400	Compressive
PO	6,500	5,400	11,900	Compressive
ML	1,600	1,500	3,100	Tensile
ON	3,500	3,000	6,500	Tensile
QP	10,600	4,500	15,100	Tensile
FQ	16,000	6,600	22,600	Compressive

The values in this table represent the stress in round numbers, on the various members in the truss, as obtained from the dead-load and wind diagrams.

The size of the timber and tension rods in the truss may now be calculated.

22. Trusses With One End Free.—On account of expansion, due to changes in temperature, in steel trusses of long spans, one end of the truss is often placed on rollers so as to allow for this expansion. This prevents the truss from pushing out the walls of the building, as would be the case if the truss were rigidly fixed at both ends. Of course, it is necessary that at least one end of the truss should be made fast, otherwise, in a wind storm, the entire roof might slide off the walls. The expansion and contraction of a steel truss is comparatively small, but there is very little elasticity to masonry and it is necessary to remember this feature when building over long spans. Now, the end of the truss that is on rollers, known as the *free end*, cannot take any lateral thrust and for this reason the reaction at this end must be vertical. Therefore, all the horizontal thrust due to the wind must be resisted by the fixed end. Of course, since the dead load acts vertically, the analysis of the stresses in a truss with one end free, for such a load, is exactly the same as in any other truss, since both of the reactions are also vertical. Therefore, it is only in the case of the wind diagram that a different method of analysis must be employed. There are two cases to consider, one

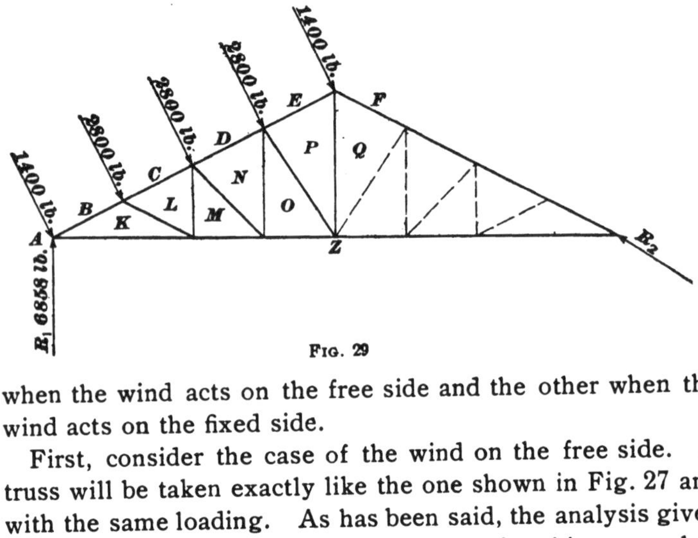

FIG. 29

when the wind acts on the free side and the other when the wind acts on the fixed side.

First, consider the case of the wind on the free side. A truss will be taken exactly like the one shown in Fig. 27 and with the same loading. As has been said, the analysis given in Fig. 26 for the dead load holds good for this truss also. The truss is redrawn in Fig. 29 and the analysis of the stresses is shown in Fig. 30. In this case, it is known that R_1 is vertical, but its magnitude is not known; both the direction and the magnitude of R_2 are unknown. To find the value of R_1, since all the forces acting on the truss are in equilibrium, moments may be taken about R_2, or the fixed end of the frame. The loads due to the wind all exert negative moments. The arm of each of these forces can be readily found from the dimensions given on Fig. 27. Thus, the arm of EF is 26.5 feet, the arm of DE is $26.5 + 11.18 = 37.68$ feet, and soon. Adding these negative moments together, it is

$$\begin{aligned} 1400 \times 26.50 &= 37100 \text{ foot-pounds} \\ 2800 \times 37.68 &= 105504 \text{ foot-pounds} \\ 2800 \times 48.86 &= 136808 \text{ foot-pounds} \\ 2800 \times 60.04 &= 168112 \text{ foot-pounds} \\ 1400 \times 72.22 &= \underline{101108} \text{ foot-pounds} \\ & 548632 \text{ foot-pounds} \end{aligned}$$

Now, the arm of R_1 about the fixed end is 80 feet. Therefore, the value of R_1 is $548{,}632 \div 80 = 6{,}858$ pounds.

The stress diagram can now be laid out. Start, for instance, at the point Z and move around the figure in

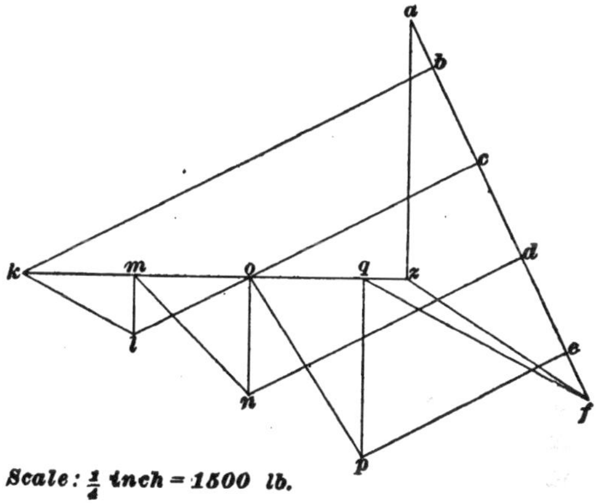

FIG. 30

the direction of the hands of a clock. Lay off, in Fig. 30, za vertical and upwards and equal to the reaction ZA, to the scale of, say, 1,500 pounds to $\frac{1}{4}$ inch. From a lay off ab equal in magnitude and direction to AB in Fig. 29; then, bc equal to BC, and so on to ef. Then, the only force that remains to be drawn is FZ or R_2, and, as this is the force that holds the truss in equilibrium, it closes the figure. The line from f to z will therefore represent R_2, both in magnitude and in direction. The direction of R_2 can now be drawn in Fig. 29, and we are in a position to analyze the stresses in the different members of the truss. First, take the joint *ABKZ*. From *b*, Fig. 30, draw *bk* parallel to *BK*; and from *z* draw *kz* parallel to *KZ*, and so on. The only difference between this

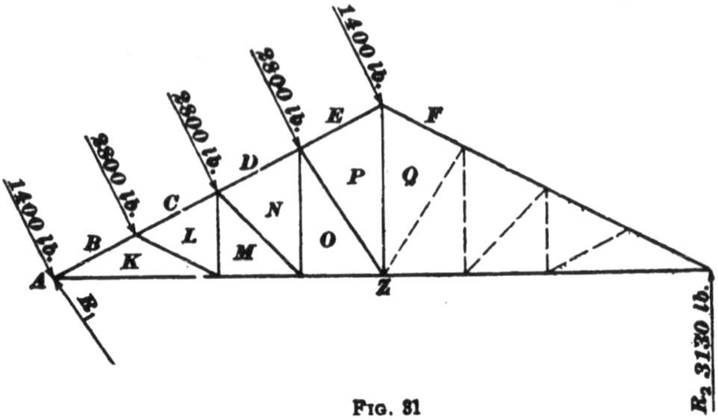

FIG. 31

analysis and the one when both ends of the truss are fixed, as shown in Fig. 28, is that *z* is not on the line *af*.

The other case is when the wind blows on the fixed side. The truss is again redrawn in Fig. 31, and the stress diagram laid out in Fig. 32.

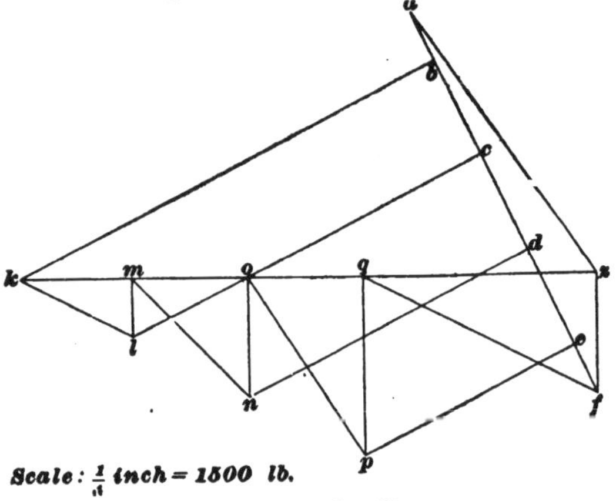

FIG. 32

In this case also, moments are taken about the fixed end. The sum of the moments of the wind loads has already been found in Art. 20 to be 250,432 foot-pounds. This quantity divided by 80, the span of the truss, gives the value of R_2, which is $R_2 = 250{,}432 \div 80 = 3{,}130$ **pounds.** Lay off the wind loads in Fig. 32 to their correct direction and scale from *a* to *f*. From *f*, draw *f z* equal in magnitude and direction to R_2. Draw *z a*; then *z a* is equal in magnitude and direction to R_1 and can be so laid off in Fig. 31. The stress diagram can now be completed. From *b*, draw *b k* parallel to *B K* in Fig. 31. From *z*, draw *k z* parallel to *K Z*, Fig. 31. Proceed to find the other stresses in the truss members in the regular way. It will again be noted that *z* is not on the line *a b*, but is now on the side opposite to where it was in Fig. 30.

The rule to follow in solving such problems as the foregoing is this: First, find the reaction at the free end by taking moments about the fixed end. Lay off the wind loads and the reaction at the free end in their proper order and direction, all to one scale. By joining the two ends of this diagram (as *z* and *a*, Fig. 32) the magnitude and direction of the reaction at the fixed end is found. Then proceed as in any other stress diagram.

DETERMINATION OF STRESSES IN THE FINK TRUSS

23. The **Polonceau,** or **Fink, truss,** shown in Fig. 33 (*a*), is a favorite form of truss. It is often built of

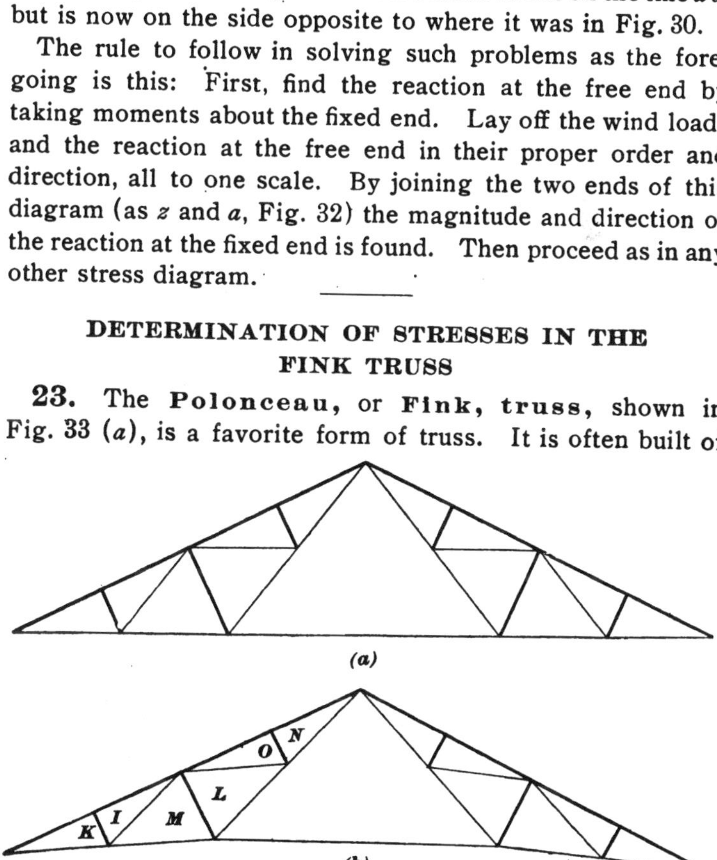

(*b*)

Fig. 33

rolled-steel sections; it is also built with wooden rafter

members, structural-steel struts, and wrought-iron tension members. Frequently, the lower chord of the truss is **cambered** (raised at the center), as shown in Fig. 33 (*b*). Cambering the lower chord in this manner gives greater headroom under the truss at the center, and somewhat improves its appearance; but it increases the stresses on all the members except *KI*, *ML*, and *ON*.

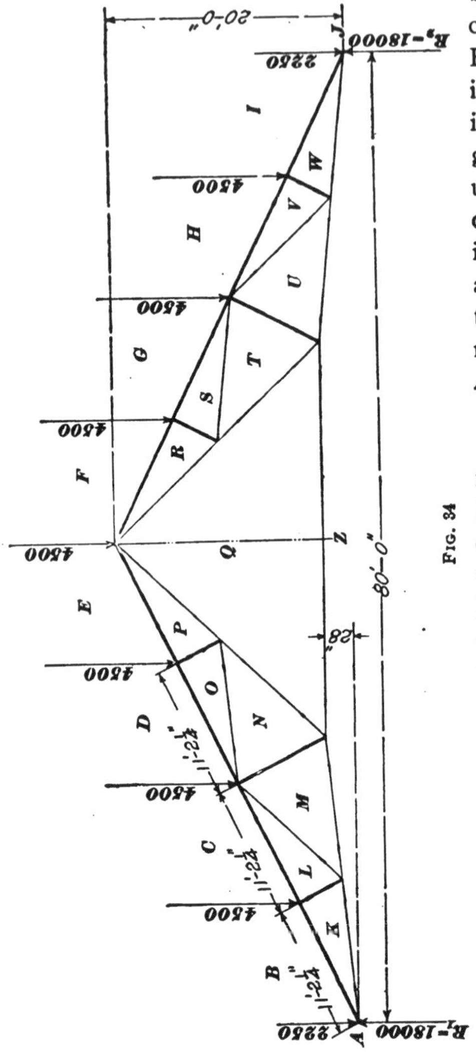

FIG. 34

24. Diagram for Vertical Loads. Fig. 34 is a frame diagram showing the vertical loads on a Fink truss, the span of which is 80 feet, the lower chord, or tie, of the truss being cambered from the horizontal 28 inches. The stress diagram for the vertical loads may be drawn, as shown in Fig. 35, by first drawing the vertical load line *a f*, and laying off to some convenient scale the loads *a b*, *b c*, *c d*, *d e*, and *e f*, designated in the frame diagram, Fig. 34, by *A B*, *B C*, *C D*, *D E*, and *E F*. Since the truss is symmetrically loaded, only one-half of the stress diagram need be drawn as the other side will be exactly the same, and consequently the loads only as far as *e f* need be laid off on the vertical load line. The reactions R_1 and R_2 are each equal to one-half of the total load, or 18,000 pounds; hence, the point *z* is located midway between *e* and *f*, and *z a* represents the reaction R_1, 18,000 pounds.

The stresses around the joint *A B K Z* may be drawn in the stress diagram by commencing at *b* and drawing *b k* parallel with *B K*, and from *z*, a line parallel with *K Z*, intersecting the first line at *k*. The polygon of forces around this joint will then be from *a* to *b*, from *b* to *k*, from *k* to *z*, and from *z* back again to *a*, the starting point. From the direction of these forces, the kind of stress on the member is observed.

The next joint in the truss to analyze is *B C L K*. In the stress diagram begin at *c* and draw *c l* parallel with *C L* in the frame diagram, then from *k*, draw a line parallel with *L K* in the frame diagram, until it intersects the line drawn from *c* at the point *l*. The polygon of forces around the joint is from *b* to *c*, from *c* to *l*, from *l* to *k*, and from *k* back to *b*, the starting point.

Around the joint *K L M Z*, the stresses are obtained by drawing from *l* a line parallel with *L M* in the frame diagram, until it intersects the line *z k* at the point *m*. The polygon of forces around this joint is from *k* to *l*, from *l* to *m*, from *m* to *z*, and from *z* back again to *k*, the starting point.

Difficulty will be encountered on attempting to analyze either the joint *C D O N M L* or *M N Q Z*, for at each of these joints there are three unknown forces, or stresses. It is not easy, by the graphic method, to solve the stresses around a joint in a structure where there are more than two unknown forces, consequently some other method of solution must be found.

On inspecting the frame diagram, Fig. 34, it will be seen that the joint *D E* is similar to the joint *B C*; the panel load, 4,500 pounds in each case, is supported by two forces, *B K* and *K L*, *D O* and *O P*, respectively. Since the directions of the forces are respectively parallel, and the panel loads are equal, the stresses in KL and OP are equal, these stresses being due only to a component of their respective panel loads. In a similar manner, it can be shown that the stresses in KL and OP are each held in equilibrium by the pairs of forces, KZ and LM, ON and PQ, and that the stresses in LM and NO are produced by equal components of the equal stresses in KL and OP; therefore, the stresses in LM and NO are equal. This gives the magnitude of the stress ON, its equal LM having already been

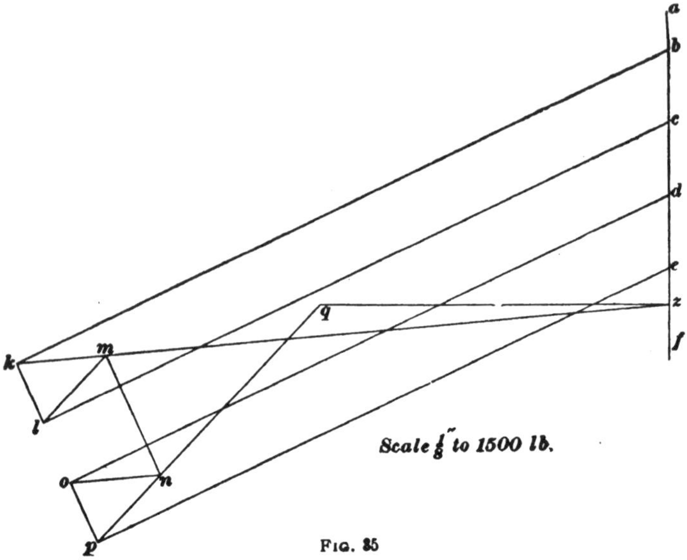

FIG. 35

determined, and the diagram can be completed as follows: Draw mn parallel to MN of the frame diagram, then draw do parallel with DO, and of such a length that when on is drawn, from its extremity o, the point n will meet the line mn midway between the lines do and ep. This construction makes the length on equal to lm, which is in accordance with the condition that the stresses in LM and ON are equal. Then the polygon of forces around this joint will be from c to d, from d to o, from o to n, from n to m, from m to l, and from l to c, the starting point.

The remaining joints, when taken in their usual order, offer no difficulty, and the other half of the diagram need not be drawn unless it is desired to check the half of the diagram just completed.

The polygon of forces that has just been traced around the joint *CDONML* affords a good illustration of the rule that the forces that meet at a joint must make a closed polygon in the stress diagram.

One of the peculiarities of the stress diagram, Fig. 35, and one that is worthy of note, as it will materially assist in drawing the diagram, is that the triangles *lkm* and *pon* are equal to each other and similar to the larger one whose base is *mn*.

25. The wind-stress diagram may now be drawn. First, the frame diagram is redrawn, and on it, as shown in Fig. 36, are designated the wind loads acting at the several joints of the truss, in a direction normal to the slope of the roof.

The reactions R_1 and R_2 may be calculated by the principle of moments. Since the truss is securely fastened at both ends, neither end being free to move in a lateral direction, these reactions will be parallel with the action of the wind on the roof, that is, normal to the slope. The reaction R_2 may first be obtained by extending its line of direction until it intersects the extension of the left-hand rafter member at the point *a'*. Then, by taking the center of moments at the left-hand reaction R_1, the magnitude of the reaction R_2 may be computed.

For convenience, reduce to feet and decimals the distance from each panel point to the point of rotation R_1; the moments about this point will then be as follows:

$5{,}625 \times 11.188 =$	62932.50	foot-pounds
$5{,}625 \times 22.375 =$	125859.38	foot-pounds
$5{,}625 \times 33.563 =$	188791.88	foot-pounds
$2{,}813 \times 44.750 =$	125881.75	foot-pounds
Total,	503465.51	foot-pounds

The distance of the center of moments of the line of action of the reaction R_2 is $44.75 + 27 = 71.75$ feet, and

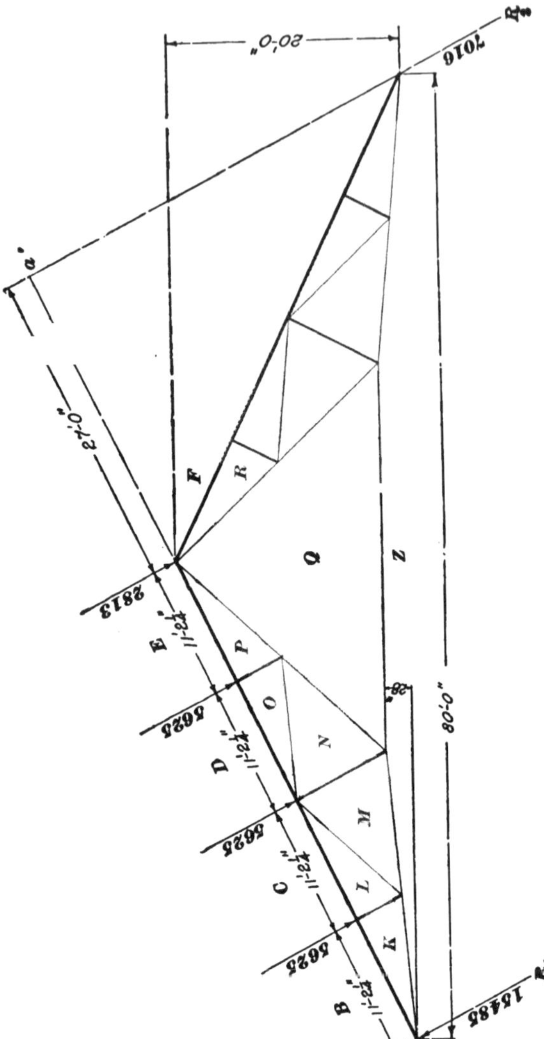
20'-0"
7016
a'
27'-0"
2813
E
11'-2¼"
5625
D
11'-2¼"
5625
C
11'-2¼"
5625
B
11'-2¼"
2813
A
15485
F
R
Q
P
O
N
M
L
K
Z
80'-0"

FIG. 36

$503,465.51 \div 71.75 = 7,016$, the magnitude of the reaction R_2, in pounds.

Since the sum of the wind loads is 22,501 pounds, which is equal to the sum of the reactions, the reaction due to the wind at R_1 is $22,501 - 7,016 = 15,485$ pounds.

The wind-stress diagram, Fig. 37, may now be drawn. First draw the load line *a* to *f* parallel to the reactions and direction of the wind loads at the several joints. Then lay off on the load line the loads *a b*, *b c*, *c d*, *d e*, and *e f*,

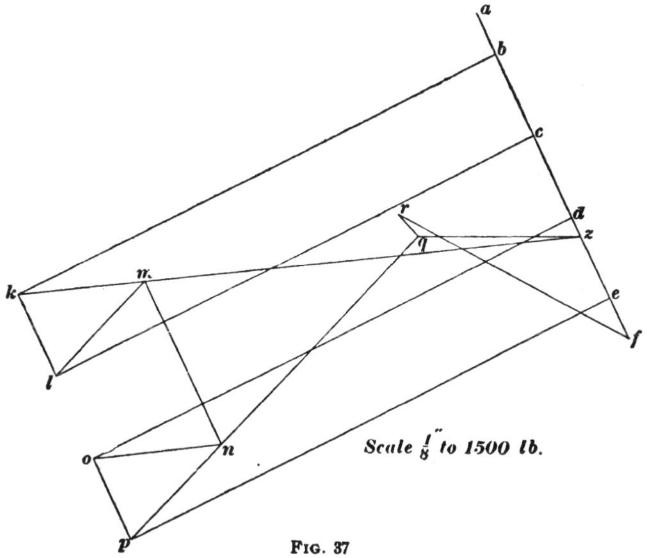

FIG. 37

which are respectively equal to the corresponding loads *A B*, *B C*, *C D*, *D E*, and *E F* in the frame diagram. Having located the point *f*, the magnitude of the reaction R_2, represented by *f z*, may be laid off upwards (the direction in which it acts), and the point *z* is thus located; then by scaling *z a*, it should be found equal to 15,485 pounds, the reaction R_1.

The polygon of external forces will then be from *a* to *b*, from *b* to *c*, from *c* to *d*, from *d* to *e*, from *e* to *f*, from *f* to *z*, and from *z* to *a*, the starting point.

The joint *CDONML* may be solved in this stress diagram in the same manner as it was solved in the vertical-load stress diagram.

The analysis of the stresses around the last joint *EFRQP* in the frame diagram is interesting from the fact that the stress *q r* closes the diagram, and, if the diagram is correctly drawn, it must be parallel to the member *QR*.

Having drawn both the wind and the vertical-load stress diagrams, the stresses in the several members in the truss may be obtained by scaling, and their magnitudes may be tabulated as follows, compression in this table being indicated by the plus sign and tension by the minus sign:

Member	Stress Due to Vertical or Dead Load Pounds	Stress Due to Wind Load Pounds	Total Stress Pounds
BK	+ 46,000	+ 34,500	+ 80,500
CL	+ 44,000	+ 34,500	+ 78,500
DO	+ 42,000	+ 34,500	+ 76,500
EP	+ 40,000	+ 34,500	+ 74,500
KL	+ 4,000	+ 5,500	+ 9,500
MN	+ 8,000	+ 11,250	+ 19,250
OP	+ 4,000	+ 5,500	+ 9,500
ZK	− 41,500	− 36,750	− 78,250
ZM	− 35,500	− 28,500	− 64,000
ZQ	− 21,000	− 11,000	− 32,000
QN	− 15,500	− 18,500	− 34,000
QP	− 21,500	− 26,000	− 47,500
FR		+ 17,000	
RQ		− 2,000	
NO	− 5,750	− 8,000	− 13,750
LM	− 5,750	− 8,000	− 13,750

STRENGTH OF RIVETS AND PINS

26. Rivets and pins are the elements by which the different sections and members of a steel truss are bound, or tied together. Pins also are sometimes used in the better class of timber construction, in which case, however, the tension members are usually made of steel.

Where plates or rolled shapes are joined by either pins or rivets, as in Fig. 38, there is more or less friction between the several parts, which acts to prevent them from being pulled apart. This is especially true when rivets are driven close against the plates while hot; in cooling, they contract between the heads and bind the plates tightly together.

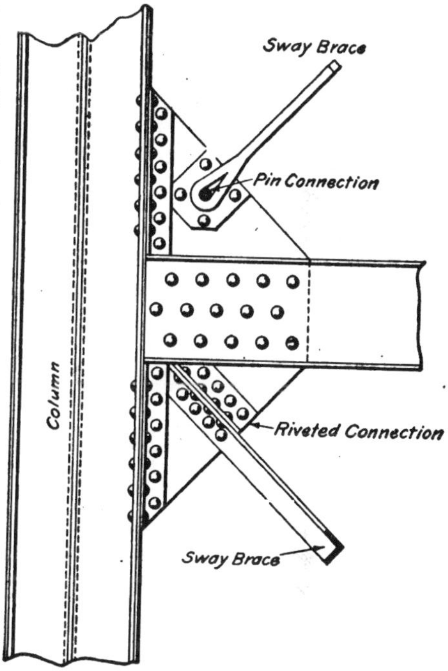

FIG. 38

The friction between plates bound together by rivets and pins, however, is not taken into account by American engineers when calculating the strength of the connections.

27. Methods of Failure.—Riveted joints may fail either by the shearing of the rivet or the crimping of the plates that they connect. When a rivet shears, the tendency is to cut straight through it across its section, as shown in

Fig. 39 (*a*) and (*b*). Where there are only two plates connected, as shown at (*a*), the tendency is to cut the rivet on the single plane *a b*. A rivet in this position is said to be in

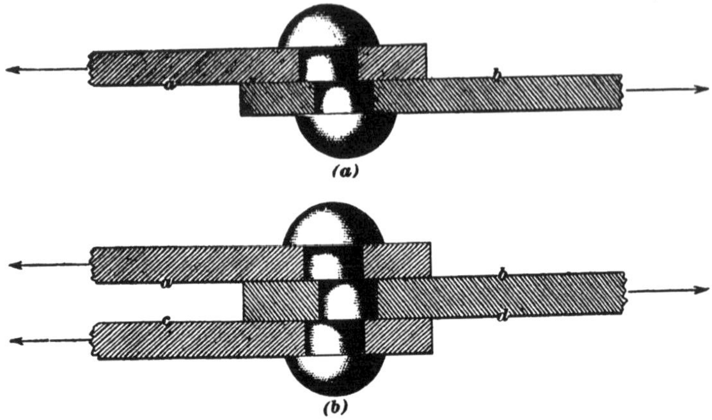

FIG. 39

single shear. At (*b*), the tendency is to cut through the rivet on both the planes *a b* and *c d*; under these conditions the rivet is in **double shear,** and it is evident that, since the rivet will shear across at two places, it will be twice as strong as where the tendency is to shear through only one section.

FIG. 40

When the diameter of the rivet is large in proportion to the thickness of the plate, the joint or connection is liable to fail by the crimping of the plate, as shown in Fig. 40. This occurs when the resistance of the plate to crushing or crimping is less than the resistance of the rivet to shear.

28. Bearing Value of Rivets.—The condition of the plates in the joint shown in Fig. 41 (*a*) is called **ordinary bearing,** while the plate *m* connected as shown at (*b*) is said to be in **web bearing.** This distinction is important, because the bearing value of a web-plate is greater than that of an outside plate, the value for web bearing being about one-third greater than for ordinary bearing.

As the tensile strength of iron and steel used in the manufacture of rivets, pins, and plates for structural work is more easily determined by tests, and, therefore, better known than either its compressive or shearing strength, it is customary to use this as a basis from which to calculate the bearing value of plates and the shearing strength of rivets and pins.

Good practice assumes that the compressive strength of steel or high-test iron is about thirteen-fifteenths of its tensile strength; that is, if the safe tensile strength of the material per square inch of section is 15,000 pounds, the safe compressive strength may be taken as thirteen-fifteenths of 15,000, or 13,000 pounds.

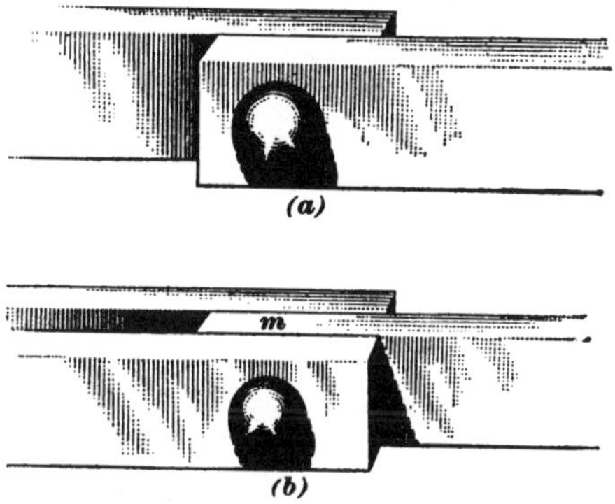

FIG 41

The shearing strength is considered to be five-sixths of the compressive strength. For example, if, as stated, the tensile strength is 15,000 pounds and the compressive strength 13,000 pounds, the shearing strength becomes five-sixths of 13,000, or 10,833 pounds per square inch of section.

In order to determine the bearing value of the plates around a rivet hole, it is necessary to consider the bearing area of the rivet on the plate. This is always assumed to be the product of the diameter of the rivet multiplied by the thickness of the plate. Owing, however, to the support that the material around the hole receives from the rest of the plate, it will safely sustain a pressure greater than the compressive strength of the material when not so supported. The safe bearing strength of the rivet on the plate, for ordinary bearing, is therefore assumed to be $1\frac{1}{2}$ times its compressive strength, and for web bearing the safe bearing strength is assumed as double the compressive strength of the material.

In deducting the rivet holes, to ascertain the net section of a riveted plate, the diameter of the hole is taken as $\frac{1}{8}$ inch larger than the diameter of the rivet.

EXAMPLE 1.—Two pieces of structural steel are joined by rivets, as shown in Fig. 42. If the tensile strength of the steel is 60,000 pounds

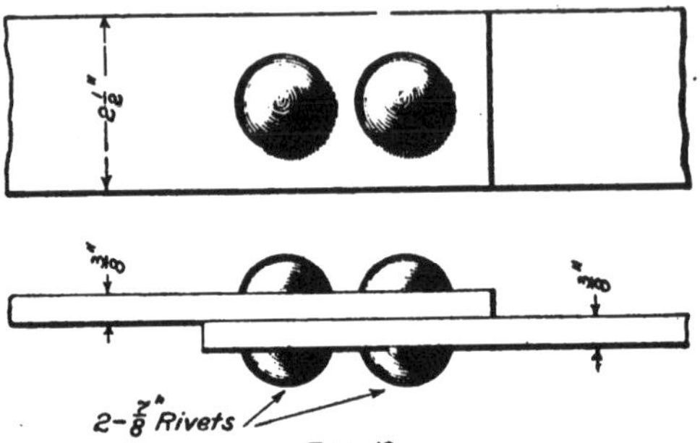

FIG. 42

per square inch and the factor of safety of 4 is used, what is the safe strength of this joint?

SOLUTION.—The safe tensile strength of the steel is 60,000 ÷ 4 = 15,000 lb. per sq. in. The width of the pieces connected is $2\frac{1}{2}$ in., from which is to be deducted 1 in. for the rivet hole, leaving a net width of $1\frac{1}{2}$ in., which, multiplied by the thickness of the plate, gives a net area of $1\frac{1}{2} \times \frac{3}{8} = .5625$ sq. in. Then, $.5625 \times 15{,}000 = 8{,}437$ lb., the strength of the plate.

Now, to determine whether the strength of the rivets is equal to the net section of the plate: Taking the compressive value of the plate as thirteen-fifteenths of 15,000 lb., or 13,000 lb. per sq. in., and the rivets being in ordinary bearing, the safe bearing strength is $13{,}000 \times 1\frac{1}{2} = 19{,}500$ lb. per sq. in. of bearing area. The bearing area is $\frac{7}{8} \times \frac{3}{8} = .328$ sq. in.; therefore, the safe bearing strength for one rivet is $19{,}500 \times .328 = 6{,}396$ lb., and for the two it is $2 \times 6{,}396 - 12{,}792$ lb.

The next point to consider is the resistance of the rivets to shear: The shearing strength of the steel is five-sixths of 13,000 = 10,833 lb. per sq. in. The area of a $\frac{7}{8}$-in. rivet is .601 sq. in., which, multiplied by 10,833, gives 6,510 lb., the shearing strength of one rivet. The total resistance to shear of the rivets in the joint is therefore $6{,}510 \times 2 = 13{,}020$ lb.

The safe resistance of the three elements entering into the strength of the joint is therefore as follows: Resistance of net section of the plate = 8,437 lb.; bearing value of the plate = 12,792 lb.; shearing strength of the two rivets = 13,020 lb.; from which it is easily seen that the strength of the joint is that of the net section of the plate, 8,437 lb. Ans.

Since the bearing value of the plate and the shearing strength of the rivets are considerably in excess of this amount, it appears that the rivets are large for the joint, and it is probable that $\frac{3}{4}$-in. rivets will give better results.

EXAMPLE 2.—One of the tension members in a structure is connected as shown in Fig. 43. The tension bars are made of structural

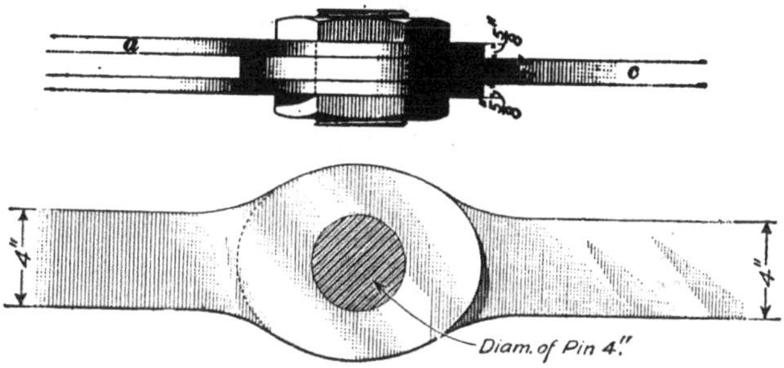

FIG. 43

steel with a safe tensile strength of 15,000 pounds per square inch. (*a*) What is the bearing value of the bar *c*? (*b*) What is the bearing value of the two bars *a*?

SOLUTION.—(*a*) The safe compressive strength of the material is thirteen-fifteenths of 15,000 lb., or 13,000 lb. Then the bearing value of the bar *c*, which may be considered as a web, is 2 × 13,000 = 26,000 lb. per sq. in. The bearing area of the pin in the bar *c* is 4 × 1 = 4 sq. in.; therefore, the bearing strength of the bar is

$$26{,}000 \times 4 = 104{,}000 \text{ lb. Ans.}$$

(*b*) As the piece *a* is in ordinary bearing, its bearing value is 13,000 × $1\frac{1}{2}$ = 19,500 lb. per sq. in. The bearing area of the two bars is $2 \times 4 \times \frac{5}{8} = 5$ sq. in.; their combined bearing strength is therefore

$$19{,}500 \times 5 = 97{,}500 \text{ lb. Ans.}$$

EXAMPLE 3.—Determine the safe strength of the riveted joint shown in Fig. 44, in which the plates and rivets each have a safe tensile strength of 16,000 pounds per square inch.

SOLUTION.—The safe tensile strength of the material being 16,000 lb., the safe compressive strength is thirteen-fifteenths of 16,000 lb., or 13,867 lb., and the shearing strength of the rivets is five-sixths of 13,867, or 11,556 lb. per sq. in. of section. The area of the section of a

$\frac{7}{8}$-in. rivet is .601 sq. in.; therefore, the total shearing strength of the three rivets, each of which is in double shear, is

$$2 \times .601 \times 11{,}556 \times 3 = 41{,}670 \text{ lb.}$$

The two outside plates are in ordinary bearing, and their bearing value is $1\frac{1}{2} \times 13{,}866 = 20{,}800$ lb. per sq. in. There are three rivet holes in each plate, each with a bearing area of $\frac{3}{8} \times \frac{7}{8} = .328$ sq. in.; the total bearing strength of the two plates is therefore

$$20{,}800 \times .328 \times 3 \times 2 = 40{,}934 \text{ lb.}$$

The bearing value of the central, or web-bearing, plate at one rivet hole is $2 \times 13{,}867$ lb. $= 27{,}734$ lb. per sq. in., and the bearing area is

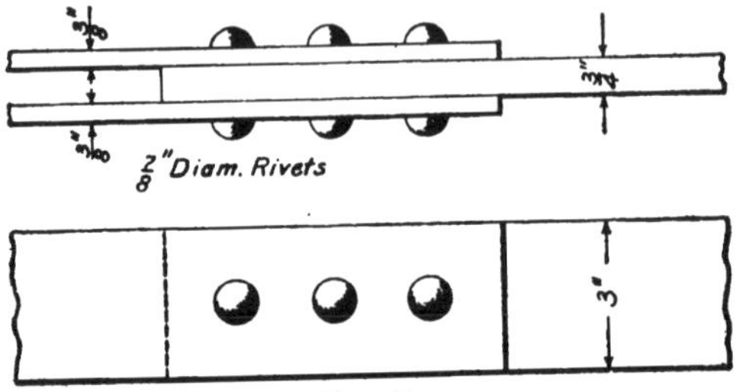

FIG. 44

$\frac{3}{4} \times \frac{7}{8} = .656$ sq. in.; the total bearing strength for the three rivets is therefore

$$27{,}734 \times .656 \times 3 = 54{,}580 \text{ lb.}$$

The safe tensile strength of the central plate is equal to its net section multiplied by 16,000, the safe unit tensile strength of the material. The net width of the plate is $3 - 1 = 2$ in., and its net area, $2 \times \frac{3}{4} = 1.5$ sq. in.; therefore, the safe strength is $1.5 \times 16{,}000 = 24{,}000$ lb. The strength of the two outside plates, calculated in the same manner, is found to be 24,000 lb. also; therefore, it is evident that, since the strength of the net section of the plate is much less than either the strength of the rivets or the bearing value of the plates, it determines the strength of the joint, which is therefore 24,000 lb. Ans.

29. Table of Bearing Values of Rivets.—In order to avoid the necessity of calculating the shearing value of the rivets and the bearing value of the riveted plates and rolled sections, Table III, showing the values of rivets, has been prepared. It will be noticed that the areas and shearing values for both double and single shear are given for rivets from $\frac{1}{2}$ inch to $\frac{7}{8}$ inch in diameter.

The least distance that rivets may be placed from the end or side of a plate is given in the third and fourth vertical columns from the left-hand side of the table. By referring to Fig. 45, the terms *end distance* and *side distance* used in the table will be readily understood.

The minimum allowable distance from the end and side of plate, as given in the table, is readily ascertained by the following: The end distance equals the thickness of the plate plus one-half the diameter of the rivet plus $\frac{1}{8}$ inch, while the side distance equals one-half the thickness of the plate plus one-half the diameter of the rivet plus $\frac{1}{8}$ inch.

The values under the heading Allowable Stress per Square Inch on Soft Steel refer to the tensile strength of the

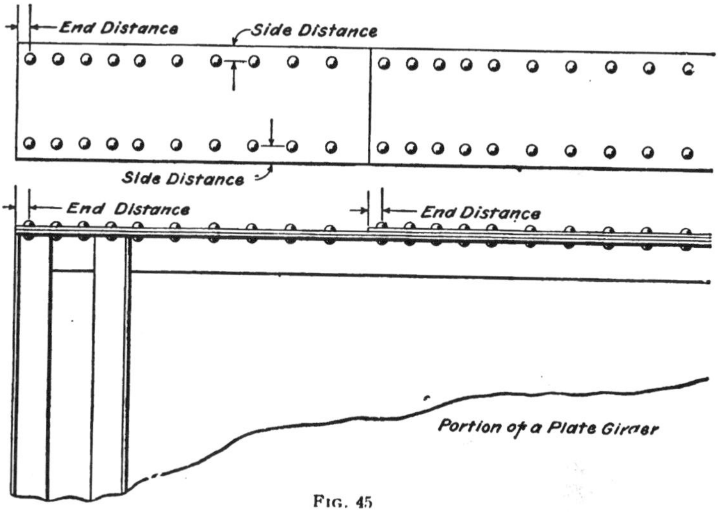

FIG. 45

material. If an allowable stress of 15,000 pounds is desired, use the value of the plates and rivets given under this head. Although not always done, the allowable tensile stress when calculating rivet values should be kept below 15,000 pounds. A stress of 10,000 or 12,000 pounds can be generally used.

If the requirements of the structure are such that values other than those given in the table are desired, the required values may be obtained by finding the sum or the difference of two of the given columns. For example, a stress of 11,000 pounds per square inch is to be allowed, and the ordinary bearing value of a $\frac{1}{4}$-inch plate at a $\frac{5}{8}$-inch rivet hole is

TABLE III

VALUES OF RIVETS

Rivet Section		Minimum Distance From Edge of Plate		Ordinary Bearing					Web Bearing				
				Plate Thickness	Allowable Stress per Square Inch on Soft Steel				Plate Thickness	Allowable Stress per Square Inch on Soft Steel			
Diameter Inch	Area Square Inch	End Inches	Side Inch		1,000	10,000	12,000	15,000		1,000	10,000	12,000	15,000
		$\frac{7}{8}$	$\frac{7}{16}$	$\frac{1}{8}$-inch plate	81.25	813	975	1,219	$\frac{1}{8}$-inch plate	108.34	1,083	1,300	1,625
		$\frac{15}{16}$	$\frac{15}{32}$	$\frac{3}{16}$-inch plate	121.28	1,219	1,463	1,828	$\frac{3}{16}$-inch plate	162.50	1,625	1,950	2,438
$\frac{1}{2}$	.196	1	$\frac{1}{2}$						$\frac{1}{4}$-inch plate	216.67	2,167	2,600	3,250
		$1\frac{1}{16}$	$\frac{17}{32}$						$\frac{5}{16}$-inch plate	270.84	2,708	3,250	4,063
				Single shear	141.56	1,416	1,699	2,123	Double shear	283.12	2,831	3,397	4,247
		$\frac{15}{16}$	$\frac{1}{2}$	$\frac{1}{8}$-inch plate	101.56	1,016	1,219	1,523	$\frac{1}{8}$-inch plate	135.42	1,354	1,625	2,031
		1	$\frac{17}{32}$	$\frac{3}{16}$-inch plate	152.34	1,523	1,823	2,285	$\frac{3}{16}$-inch plate	203.13	2,031	2,438	3,047
$\frac{5}{8}$	.307	$1\frac{1}{16}$	$\frac{9}{16}$	$\frac{1}{4}$-inch plate	203.13	2,031	2,437	3,046	$\frac{1}{4}$-inch plate	270.84	2,708	3,250	4,063
		$1\frac{1}{8}$	$\frac{19}{32}$						$\frac{5}{16}$-inch plate	338.55	3,386	4,063	5,078
		$1\frac{3}{16}$	$\frac{5}{8}$						$\frac{3}{8}$-inch plate	406.26	4,063	4,875	6,094
				Single shear	221.73	2,217	2,661	3,326	Double shear	443.46	4,435	5,322	6,652

$\frac{3}{4}$	.442	1	$\frac{9}{16}$	$\frac{1}{8}$-inch plate	121.88	1,219	1,463	1,828	$\frac{1}{8}$-inch plate	162.50	1,625	1,950	2,438
		$1\frac{1}{16}$	$\frac{19}{32}$	$\frac{3}{16}$-inch plate	182.81	1,828	2,193	2,742	$\frac{3}{16}$-inch plate	243.75	2,438	2,925	3,656
		$1\frac{1}{8}$	$\frac{5}{8}$	$\frac{1}{4}$-inch plate	243.75	2,438	2,925	3,656	$\frac{1}{4}$-inch plate	325.00	3,250	3,900	4,875
		$1\frac{3}{16}$	$\frac{21}{32}$	$\frac{5}{16}$-inch plate	304.69	3,047	3,656	4,570	$\frac{5}{16}$-inch plate	406.25	4,063	4,875	6094
		$1\frac{1}{4}$	$\frac{11}{16}$						$\frac{3}{8}$-inch plate	487.50	4,875	5,850	7,313
		$1\frac{5}{16}$	$\frac{23}{32}$						$\frac{7}{16}$-inch plate	568.75	5,688	6,825	8,531
				Single shear	319.24	3,192	3,830	4,788	Double shear	638.47	6,385	7,662	9,577
$\frac{7}{8}$	.601	$1\frac{1}{16}$	$\frac{5}{8}$	$\frac{1}{8}$-inch plate	142.19	1,422	1,706	2,133	$\frac{1}{8}$-inch plate	189.58	1,896	2,275	2844
		$1\frac{1}{8}$	$\frac{21}{32}$	$\frac{3}{16}$-inch plate	213.28	2,133	2,559	3,199	$\frac{3}{16}$-inch plate	284.37	2,844	3,413	4,266
		$1\frac{3}{16}$	$\frac{11}{16}$	$\frac{1}{4}$-inch plate	284.37	2,844	3,413	4,266	$\frac{1}{4}$-inch plate	379.17	3,792	4,550	5,687
		$1\frac{1}{4}$	$\frac{23}{32}$	$\frac{5}{16}$-inch plate	355.47	3,555	4,266	5,332	$\frac{5}{16}$-inch plate	473.96	4,740	5,687	7,109
		$1\frac{5}{16}$	$\frac{3}{4}$	$\frac{3}{8}$-inch plate	426.57	4,266	5,119	6,398	$\frac{3}{8}$-inch plate	568.75	5,688	6,825	8,531
		$1\frac{3}{8}$	$\frac{25}{32}$						$\frac{7}{16}$-inch plate	663.54	6,635	7,962	9,953
		$1\frac{7}{16}$	$\frac{13}{16}$						$\frac{1}{2}$-inch plate	758.33	7,583	9,100	11,375
		$1\frac{1}{2}$	$\frac{27}{32}$						$\frac{9}{16}$-inch plate	853.12	8,531	10,237	12,796
				Single shear	434.07	4,341	5,209	6,511	Double shear	868.15	8,682	10,418	1 [illegible]

NOTE.—Ordinary bearing = $1\frac{1}{4}$ × unit compressive stress × bearing area. Web bearing = 2 × unit compressive stress × bearing area. Shear = $\frac{3}{4}$ × unit compressive stress × section. Compression = [illegible] × allowable tensile stress.

TABLE III

VALUES OF RIVETS

Rivet Section		Minimum Distance From Edge of Plate		Ordinary Bearing					Web Bearing				
				Plate Thickness	Allowable Stress per Square Inch on Soft Steel				Plate Thickness	Allowable Stress per Square Inch on Soft Steel			
Diameter Inch	Area Square Inch	End Inches	Side Inch		1,000	10,000	12,000	15,000		1,000	10,000	12,000	15,000
		7/8	7/16	1/8-inch plate	81.25	813	975	1,219	1/8-inch plate	108.34	1,083	1,300	1,625
		15/16	15/32	3/16-inch plate	121.28	1,219	1,463	1,828	3/16-inch plate	162.50	1,625	1,950	2,438
1/2	.196	1	1/2						1/4-inch plate	216.67	2,167	2,600	3,250
		1 1/16	17/32						5/16-inch plate	270.84	2,708	3,250	4,063
				Single shear	141.56	1,416	1,699	2,123	Double shear	283.12	2,831	3,397	4,247
		15/16	1/2	1/8-inch plate	101.56	1,016	1,219	1,523	1/8-inch plate	135.42	1,354	1,625	2,031
		1	17/32	3/16-inch plate	152.34	1,523	1,823	2,285	3/16-inch plate	203.13	2,031	2,438	3,047
		1 1/16	9/16	1/4-inch plate	203.13	2,031	2,437	3,046	1/4-inch plate	270.84	2,708	3,250	4,063
5/8	.307	1 1/8	19/32						5/16-inch plate	338.55	3,386	4,063	5,078
		1 3/16	5/8						3/8-inch plate	406.26	4,063	4,875	6,094
				Single shear	221.73	2,217	2,661	3,326	Double shear	443.46	4,435	5,322	6,652

		1	9/16	1/8-inch plate	121.88	1,219	1,463	1,828	1/8-inch plate	162.50	1,625	1,950	2,438
		1 1/16	19/32	3/16-inch plate	182.81	1,828	2,193	2,742	3/16-inch plate	243.75	2,438	2,925	3,656
		1 1/8	5/8	1/4-inch plate	243.75	2,438	2,925	3,656	1/4-inch plate	325.00	3,250	3,900	4,875
3/4	.442	1 3/16	21/32	5/16-inch plate	304.69	3,047	3,656	4,570	5/16-inch plate	406.25	4,063	4,875	6,094
		1 1/4	11/16						3/8-inch plate	487.50	4,875	5,850	7,313
		1 5/16	23/32						7/16-inch plate	568.75	5,688	6,825	8,531
				Single shear	319.24	3,192	3,830	4,788	Double shear	638.47	6,385	7,662	9,577
		1 1/16	5/8	1/8-inch plate	142.19	1,422	1,706	2,133	1/8-inch plate	189.58	1,896	2,275	2,844
		1 1/8	21/32	3/16-inch plate	213.28	2,133	2,559	3,199	3/16-inch plate	284.37	2,844	3,413	4,266
		1 3/16	11/16	1/4-inch plate	284.37	2,844	3,413	4,266	1/4-inch plate	379.17	3,792	4,550	5,687
		1 1/4	23/32	5/16-inch plate	355.47	3,555	4,266	5,332	5/16-inch plate	473.96	4,740	5,687	7,109
7/8	.601	1 5/16	3/4	3/8-inch plate	426.57	4,266	5,119	6,398	3/8-inch plate	568.75	5,688	6,825	8,531
		1 3/8	25/32						7/16-inch plate	663.54	6,635	7,962	9,953
		1 7/16	13/16						1/2-inch plate	758.33	7,583	9,100	11,375
		1 1/2	27/32						9/16-inch plate	853.12	8,531	10,237	12,796
				Single shear	434.07	4,341	5,209	6,511	Double shear	868.15	8,682	10,418	13,022

NOTE.—Ordinary bearing = 1¼ × unit compressive stress × bearing area. Web bearing = 2 × unit compressive stress × bearing area. Shear = ¾ × unit compressive stress × section. Compression = 13/16 × allowable tensile stress.

required. By adding the value given for an allowable stress of 1,000 pounds to that given for 10,000 pounds, the required bearing value may be determined; in this case it is 203.13 + 2,031 = 2,234.13, which is the allowable bearing value of a $\frac{1}{4}$-inch plate in ordinary bearing around a $\frac{5}{8}$-inch diameter rivet, where the safe tensile stress of the materials is taken at 11,000 pounds per square inch.

30. Pins Subjected to Bending Stresses.—Pins, when used to connect the several members of a structure, besides being subjected to shearing in the same manner as rivets, may be required to resist heavy bending stresses; they may then be regarded as solid cylindrical beams and

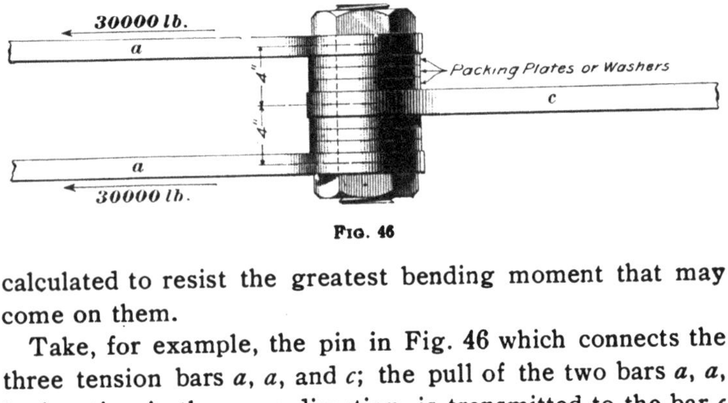

FIG. 46

calculated to resist the greatest bending moment that may come on them.

Take, for example, the pin in Fig. 46 which connects the three tension bars *a*, *a*, and *c*; the pull of the two bars *a*, *a*, both acting in the same direction, is transmitted to the bar *c*

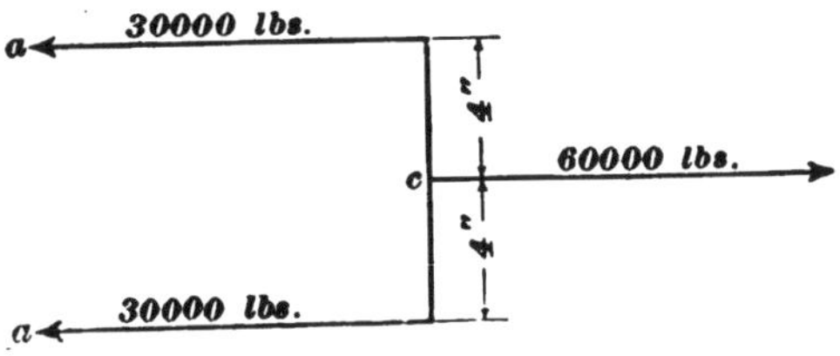

FIG. 47

by means of the pin. The stress on each bar *a* is 30,000 pounds, consequently the stress on the bar *c* must be 60,000 pounds. Assuming a maximum unit stress of 15,000 pounds per square inch, it is desired to find what diameter of pin

TABLE IV

RESISTING MOMENTS OF PINS

With Extreme Fiber Stresses Varying From 15,000 to 25,000 Pounds per Square Inch

Diameter of Pin Inches	Area of Pin Square Inches	Moments, in Inch-Pounds, for Fiber Strains of			
		15,000 lb. per Square Inch	20,000 lb. per Square Inch	22,000 lb. per Square Inch	25,000 lb. per Square Inch
1	.785	1,470	1,960	2,160	2,450
1⅛	.994	2,100	2,800	3,080	3,500
1¼	1.227	2,880	3,830	4,220	4,790
1⅜	1.485	3,830	5,100	5,620	6,380
1½	1.767	4,970	6,630	7,290	8,280
1⅝	2.074	6,320	8,430	9,270	10,500
1¾	2.405	7,890	10,500	11,570	13,200
1⅞	2.761	9,710	12,900	14,240	16,200
2	3.142	11,800	15,700	17,280	19,600
2⅛	3.547	14,100	18,800	20,730	23,600
2¼	3.976	16,800	22,400	24,600	28,000
2⅜	4.430	19,700	26,300	28,900	32,900
2½	4.909	23,000	30,700	33,700	38,400
2⅝	5.412	26,600	35,500	39,000	44,400
2¾	5.940	30,600	40,800	44,900	51,000
2⅞	6.492	35,000	46,700	51,300	58,300
3	7.069	39,800	53,000	58,300	66,300
3⅛	7.670	44,900	59,900	65,900	74,900
3¼	8.296	50,600	67,400	74,100	84,300
3⅜	8.946	56,600	75,500	83,000	94,400
3½	9.621	63,100	84,200	92,600	105,200
3⅝	10.321	70,100	93,500	102,900	116,900
3¾	11.045	77,700	103,500	113,900	129,400
3⅞	11.793	85,700	114,200	125,600	142,800

TABLE IV—(*Continued*)

Diameter of Pin Inches	Area of Pin Square Inches	Moments, in Inch-Pounds, for Fiber Strains of 15,000 lb per Square Inch	20,000 lb. per Square Inch	22,000 lb. per Square Inch	25,000 lb. per Square Inch
4	12.566	94,200	125,700	138,200	157,100
4⅛	13.364	103,400	137,800	151,600	172,300
4¼	14.186	113,000	150,700	165,800	188,400
4⅜	15.033	123,300	164,400	180,800	205,500
4½	15.904	134,200	178,900	196,800	223,700
4⅝	16.800	145,700	194,300	213,700	242,800
4¾	17.721	157,800	210,400	231,500	263,000
4⅞	18.665	170,600	227,500	250,200	284,400
5	19.635	184,100	245,500	270,000	306,800
5⅛	20.629	198,200	264,300	290,700	330,400
5¼	21.648	213,100	284,100	312,500	355,200
5⅜	22.691	228,700	304,900	335,400	381,100
5½	23.758	245,000	326,700	359,300	408,300
5⅝	24.850	262,100	349,500	384,400	436,800
5¾	25.967	280,000	373,300	410,600	466,600
5⅞	27.109	298,600	398,200	438,000	497,700
6	28.274	318,100	424,100	466,500	530,200
6⅛	29.465	338,400	451,200	496,300	564,000
6¼	30.680	359,500	479,400	527,300	599,200
6⅜	31.919	381,500	508,700	559,600	635,900
6½	33.183	404,400	539,200	593,100	674,000
6⅝	34.472	428,200	570,900	628,000	713,700
6¾	35.785	452,900	603,900	664,200	754,800
6⅞	37.122	478,500	638,000	701,800	797,500
7	38.485	505,100	673,500	740,800	841,900
7⅛	39.871	532,700	710,200	781,200	887,800
7¼	41.282	561,200	748,200	823,000	935,300
7⅜	42.718	590,700	787,600	866,300	984,500

TABLE IV—(*Continued*)

Diameter of Pin Inches	Area of Pin Square Inches	Moments, in Inch-Pounds, for Fiber Strains of			
		15,000 lb. per Square Inch	20,000 lb. per Square Inch	22,000 lb. per Square Inch	25,000 lb. per Square Inch
7½	44.179	621,300	828,400	911,200	1,035,400
7⅝	45.664	652,900	870,500	957,500	1,088,100
7¾	47.173	685,500	914,000	1,005,300	1,142,500
7⅞	48.707	719,200	958,900	1,054,800	1,198,700
8	50.265	754,000	1,005,300	1,105,800	1,256,600
8⅛	51.849	789,900	1,053,200	1,158,500	1,316,500
8¼	53.456	826,900	1,102,500	1,212,800	1,378,200
8⅜	55.088	865,100	1,153,400	1,268,800	1,441,800
8½	56.745	904,400	1,205,800	1,326,400	1,507,300
8⅝	58.426	944,900	1,259,800	1,385,800	1,574,800
8¾	60.132	986,500	1,315,400	1,446,900	1,644,200
8⅞	61.862	1,029,400	1,372,500	1,509,800	1,715,700
9	63.617	1,073,500	1,431,400	1,574,500	1,789,200
9⅛	65.397	1,118,900	1,491,900	1,641,100	1,864,800
9¼	67.201	1,165,500	1,554,000	1,709,400	1,942,500
9⅜	69.029	1,213,400	1,617,900	1,779,600	2,022,300
9½	70.882	1,262,600	1,683,400	1,851,800	2,104,300
9⅝	72.760	1,313,100	1,750,800	1,925,900	2,188,500
9¾	74.662	1,364,900	1,819,900	2,001,900	2,274,900
9⅞	76.590	1,418,100	1,890,800	2,079,900	2,363,500
10	78.540	1,472,600	1,963,500	2,159,900	2,454,400
10¼	82.520	1,585,900	2,114,500	2,325,900	2,643,100
10½	86.590	1,704,700	2,273,000	2,500,200	2,841,200
10¾	90.760	1,829,400	2,439,300	2,683,200	3,049,100
11	95.030	1,960,100	2,613,400	2,874,800	3,266,800
11¼	99.400	2,096,800	2,795,700	3,075,400	3,494,800
11½	103.870	2,239,700	2,986,300	3,284,800	3,732,800
12	113.100	2,544,700	3,392,900	3,732,200	4,241,200

is required to resist the bending moment produced by the stresses exerted on it.

In calculating the bending moment on a pin, the forces acting on it through the several members are considered as being applied at the center of the bearings. In Fig. 46, the distance between the centers of the bearings of the members is 4 inches, and by referring to the diagram, Fig. 47, it is seen that the greatest bending moment is at *c*, and is equal to 30,000 × 4 = 120,000 inch-pounds.

Having found the greatest bending moment on the pin, it is necessary to determine its diameter, in order that its resisting moment may equal the moment of the bending stresses.

By referring to *Design of Beams*, we find the section modulus of a circular section to be

$$K = \frac{AD}{8} = \frac{.7854\,D^2 \times D}{8} = .0982\,D^3$$

The allowable unit stress on the material is $S = 15{,}000$ pounds per square inch, and the bending moment is $M = 120{,}000$ inch-pounds; substituting these values in the formula for a beam, we have $M = SK$, or $120{,}000 = 15{,}000 \times .0982\,D^3$, from which we get

$$D^3 = \frac{120{,}000}{15{,}000 \times .0982} = 81.46$$

The diameter of the pin is therefore

$$D = \sqrt[3]{81.46} = 4.335 = 4\tfrac{3}{8} \text{ inches, nearly}$$

31. Table of Resisting Moments of Pins.—To avoid the necessity of calculating the resisting moment of pins, Table IV will be found convenient. This table gives the resisting moments of pins from 1 to 12 inches in diameter, calculated for allowable fiber stresses of 15,000, 20,000, 22,000, and 25,000 pounds per square inch.

32. Resultant Moment of Several Stresses.—When a pin is used at a joint at which several members, extending in different directions, meet, as shown in Fig. 48, it is necessary to combine the stresses so as to find the resultant that gives the greatest bending moment. This is conveniently done by first resolving the stresses on each of the different members into vertical and horizontal components, and calculating the bending moments produced in each of these directions by all the corresponding components. The maximum bending moment is then given by the resultant of these two bending moments.

The details of the method for finding the maximum bending stress on a pin will be made clear by a study of the following illustrative examples:

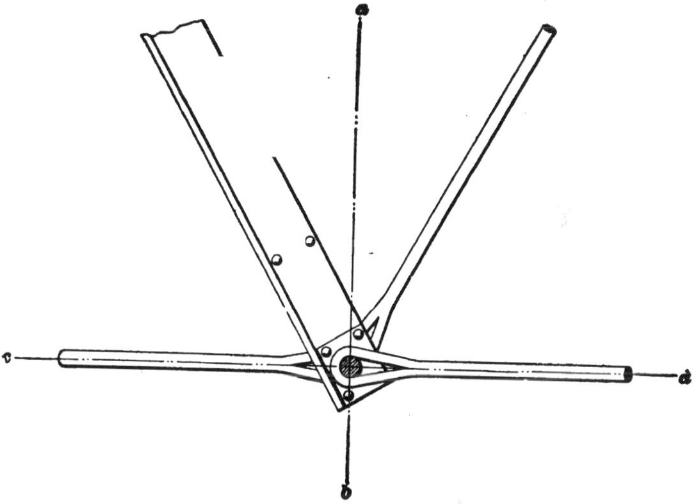

FIG. 48

Fig. 49 shows one of the lower joints of a roof truss. At this joint there are four sets of members, two of which act in a horizontal, while one acts in a vertical, direction; since they already act in the directions of the required components, these forces need not be resolved. There is, however, one inclined member in which there is a compressive stress of 40,000 pounds, and this stress must be resolved into its vertical and horizontal components. Draw the line ab parallel to the strut and of such a length as to represent the magnitude of the stress. From a, draw the horizontal line ac intersecting the vertical line at the point c. The direction of the forces around the triangle is shown by the arrows. On measuring the line *a c*, the horizontal component of the stress in *a b* is found to be 20,000 pounds, while the vertical component of the stress is found to be 34,650 pounds.

Having determined these components, a diagram showing all the horizontal stresses acting on the pin and tending to

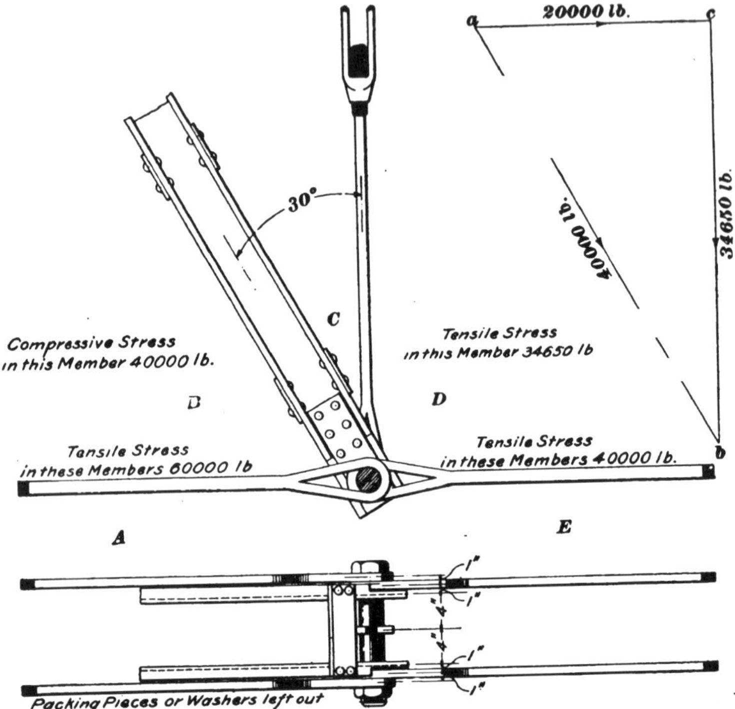

FIG. 49

bend it, and also another showing all the vertical forces, should be drawn as illustrated in Fig. 50 (*a*) and (*b*), the distance from center to center of the members being taken from the detail plan of the joint, Fig. 49. It must always be remembered that, in accordance with the principles of equilibrium, the sum of the resultants of the forces acting on the pin in any one direction must equal the sum of all the resultants acting in the opposite direction; otherwise, the pin would move in the direction of the greater sum, and the structure would fall. Thus, from Fig. 49, it is readily seen that the vertical component of the stress in BC acts in an opposite direction to the stress in the member CD, while the horizontal component acts in opposition to the stresses in the member AB, and in the same direction as the stress in DE. This makes the algebraic sum of all the components in either the horizontal or vertical direction equal to zero, and fulfils the condition of equilibrium.

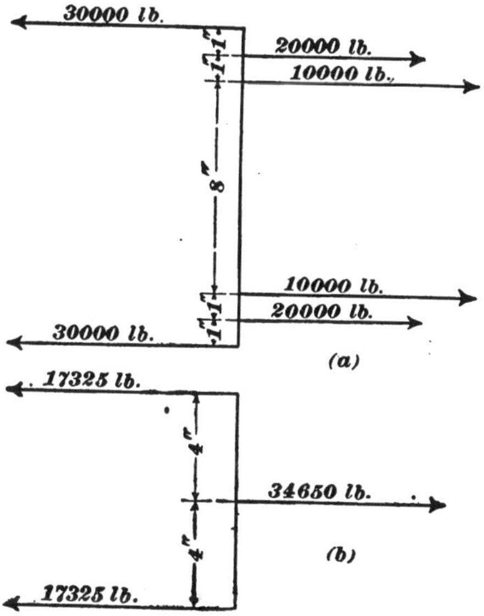

FIG. 50

The resultant of the vertical and horizontal bending moments may also be calculated by the rule for finding the length of the hypotenuse of a right triangle; for example, in this case the lengths of the sides are represented by the horizontal bending moment of 40,000 inch-pounds, and the vertical bending moment of 69,300 inch-pounds; the resultant bending moment is therefore

$$\sqrt{40,000^2 + 69,300^2} = 80,015 \text{ inch-pounds.}$$

In order to determine the required size of pin for this joint, assume a safe fiber stress of 15,000 pounds per square inch, then, by referring to Table IV, it is seen that a pin $3\frac{7}{8}$ inches in diameter, under a fiber stress of 15,000 pounds per square inch, has a resisting moment of 85,700 inch-pounds, which is very nearly the value required by the conditions.

It is necessary to remember, however, that in all cases the pin should be examined for shear as well as bending stresses.

EXAMPLES FOR PRACTICE

1. What are the safe strengths of $\frac{7}{8}$-, $\frac{3}{4}$-, and $\frac{5}{8}$-inch rivets, in double shear, and also in single shear, assuming that the safe tensile strength of the material used in their manufacture is 15,000 pounds per square inch of section?

	Diameter of Rivet Inch	Double Shear Pounds	Single Shear Pounds
Ans.	$\frac{7}{8}$	13,022	6,511
	$\frac{3}{4}$	9,577	4,788
	$\frac{5}{8}$	6,652	3,326

2. What pulling force will two pieces of $\frac{3}{8}'' \times 2\frac{1}{2}''$ bar safely resist, provided that they are connected at the end by two $\frac{3}{4}$-inch diameter rivets, as shown in Fig. 51? The safe tensile strength of the material in rivet and bar is 15,000 pounds.

Ans. 9,140 lb.

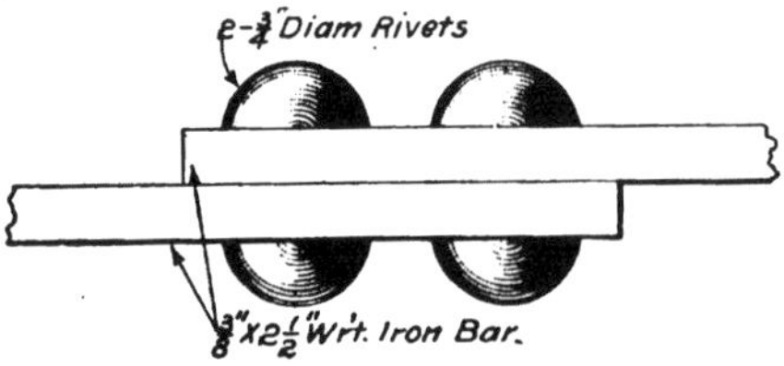

Fig. 51

3. What is the safe resisting moment of a pin 5 inches in diameter, if the safe unit strength of the material is 20,000 pounds? Ans. 245,500

4. In Fig. 52 is shown a pin connection, the pull on the tension bar *a* being 140,000 pounds. If the safe shearing strength of the material in the pin is 10,000 pounds per square inch, and the safe fiber stress in bending is 15,000 pounds per square inch: (*a*) what size of

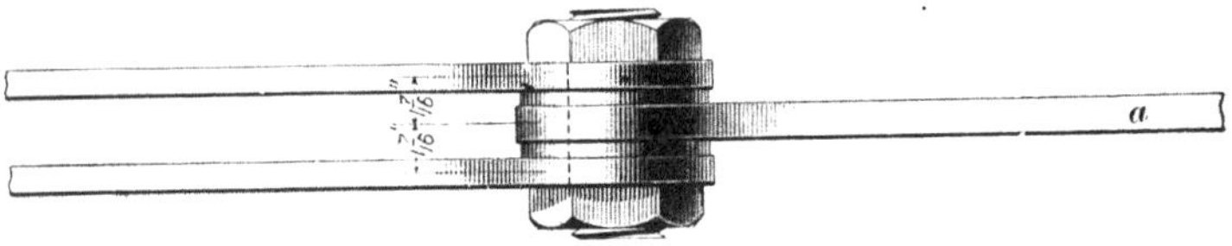

Fig. 52

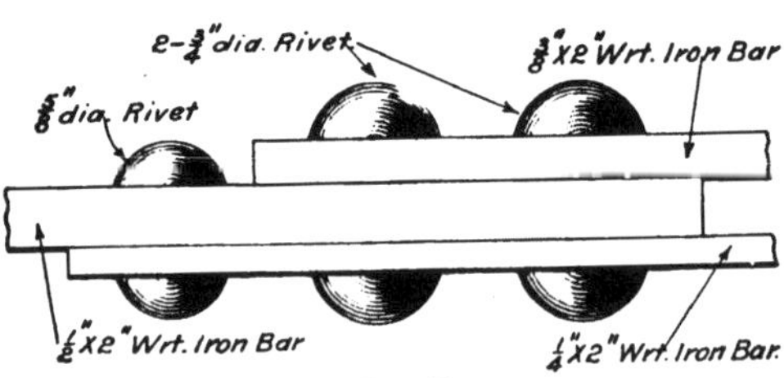

Fig. 53

pin will be required to resist the shear? (*b*) what size will be required to resist the bending?

Ans. { (*a*) 3 in. in diameter; (*b*) $4\frac{1}{8}$ in. in diameter }

5. It is necessary to construct the connection of a tension member as shown in Fig. 53. What is the safe load that this member will carry, if the safe tensile strength of the material in both the rivets and bars is 18,000 pounds per square inch? Ans. 11,250 lb.

TRUSS DESIGN

33. Designing the Members of a Truss.—Let the truss described in Art. 20 be considered, as it is a quite common type. In Art. 21 is given the stress in the various members. The first vertical tension member is ML, and has a total pull on it of 3,100 pounds. This rod will be made of soft steel, the tensile strength of which is 56,000 pounds. If a factor of safety of 4 is used, the safe load will be $56{,}000 \div 4 = 14{,}000$ pounds per square inch. A rod $\frac{3}{4}$ inch in diameter has an area at the root of the screw thread, according to Table V, of .301 square inch. Then, the safe strength of a rod $\frac{3}{4}$ inch in diameter, threaded at the ends, is $.301 \times 14{,}000 = 4{,}214$ pounds. This member being required to support 3,100 pounds only, a soft-steel rod $\frac{3}{4}$ inch in diameter is amply strong. The tension member ON is subjected to a stress of 6,500 pounds. A rod 1 inch in diameter has an area at the root of the thread, according to Table V, of .55 square inch. Hence, its safe strength is $.55 \times 14{,}000 = 7{,}700$ pounds, which is slightly in excess of that required. The member QP is subjected to a stress of 15,100 pounds. The area at the root of the thread of a rod $1\frac{1}{2}$ inches in diameter is 1.29 square inches, and the safe strength is $1.29 \times 14{,}000 = 18{,}060$ pounds, which exceeds the strength required.

To calculate the size of timber demanded for the rafter member, note that it is usual in timber trusses to make each rafter member of one size and of one length. This truss requires, then, a large stick of timber, say about 45 feet long, for the rafter members, and can no doubt be obtained

TABLE V

STANDARD SCREW THREADS AND NUTS

Diameter of Rod or Bolt	Threads per Inch	Diameter at Root of Thread	Area of Rod or Bolt at Root of Thread	Short Diameter of Nuts	Long Diameter, Hexagon Nuts	Long Diameter, Square Nuts	Thickness of Nuts
1/4	20	.185	.026	1/2	37/64	7/10	1/4
5/16	18	.240	.045	19/32	11/16	19/12	5/16
3/8	16	.294	.067	11/16	51/64	63/64	3/8
7/16	14	.344	.092	25/32	7/8	1 7/64	7/16
1/2	13	.400	.125	7/8	1	1 15/64	1/2
9/16	12	.454	.161	31/32	1 1/8	1 23/64	9/16
5/8	11	.507	.201	1 1/16	1 7/32	1 1/2	5/8
3/4	10	.620	.301	1 1/4	1 7/16	1 49/64	3/4
7/8	9	.731	.419	1 7/16	1 21/32	2 1/32	7/8
1	8	.837	.550	1 5/8	1 7/8	2 19/64	1
1 1/8	7	.940	.693	1 13/16	2 3/32	2 9/16	1 1/8
1 1/4	7	1.065	.890	2	2 5/16	2 53/64	1 1/4
1 3/8	6	1.160	1.056	2 3/16	2 17/32	3 3/32	1 3/8
1 1/2	6	1.284	1.294	2 3/8	2 3/4	3 23/64	1 1/2
1 5/8	5 1/2	1.389	1.515	2 9/16	2 31/32	3 5/8	1 5/8
1 3/4	5	1.491	1.746	2 3/4	3 3/16	3 57/64	1 3/4
1 7/8	5	1.616	2.051	2 15/16	3 13/32	4 5/32	1 7/8
2	4 1/2	1.712	2.301	3 1/8	3 5/8	4 27/64	2
2 1/4	4 1/2	1.962	3.023	3 1/2	4 1/64	4 61/64	2 1/4
2 1/2	4	2.176	3.718	3 7/8	4 1/2	5 31/64	2 1/2
2 3/4	4	2.426	4.622	4 1/4	4 29/32	6	2 3/4
3	3 1/2	2.629	5.428	4 5/8	5 3/8	6 17/32	3
3 1/4	3 1/2	2.879	6.509	5	5 13/16	7 1/16	3 1/4
3 1/2	3 1/4	3.100	7.547	5 3/8	6 7/64	7 39/64	3 1/2
3 3/4	3	3.318	8.641	5 3/4	6 21/32	8 1/8	3 3/4
4	3	3.567	9.993	6 1/8	7 3/32	8 41/64	4

by special order. As the maximum stress on the rafter, made of one piece of timber throughout, is at BK, it will be necessary to calculate its size at this point only. Assume that the truss supports heavy purlins only at the panel points, or joints, of the truss, common rafters being laid up and down the roof, resting on these purlins, as shown in Fig. 54. This concentrates all the load on the truss at the panel points, and BK is consequently not to be estimated as a beam, sustaining as it does only the compressive stress as obtained by the diagram.

The total compressive stress in BK is 39,000 pounds, and its length is 11.18 feet = 134 inches, nearly. Using an 8″ × 8″ yellow-pine timber, the ultimate compressive strength, parallel to the grain, as given in *Design of Beams*, is 5,000 pounds per square inch. By a formula given in *Design of Columns*, the ultimate compressive breaking strength of BK as a column is

$$S = U - \left(\frac{U \times L}{100\,D}\right) = 5{,}000 - \left(\frac{5{,}000 \times 134}{100 \times 8}\right)$$

$$= 5{,}000 - 838 = 4{,}162 \text{ pounds per square inch}$$

If, on account of its uncertain nature, a factor of safety of 5 is used for the timber, the safe bearing value of the column is then 4,162 ÷ 5 = 832 pounds per square inch. The area of an 8″ × 8″ column is 64 square inches, which multiplied by 832 gives a safe load of 53,248 pounds. While considerably in excess of the 39,000 pounds required, this is the nearest even-sized timber that can be used for this member.

To determine the size of timber required for the tie-member, bear in mind that the greatest pull, or tensile stress, on this member is at KZ, and amounts to 37,300 pounds. This member, made of yellow-pine timber, must be spliced, say, at the center. It is safe to assume that, in making the connection and joints, one-third of this timber will be cut away. The ultimate tensile strength of yellow pine is given in *Design of Beams* as 12,000 pounds per square inch. This divided by the factor of safety of 10 equals 1,200 pounds, the safe stress that a square inch will sustain.

For convenience in construction, a 6″ × 8″ timber is selected. The area of this timber is 48 square inches; two-thirds of this is 32 square inches. Then, 32 × 1,200 = 38,400 pounds. A 6″ × 8″ timber is thus amply strong, but on drawing the truss and laying out the detail of its connections, it may be found necessary to use a timber 8 in. × 8 in. In determining the size of the member *PO*, bear in mind that the stress on this member is 11,900 pounds. The member *PO* in the frame diagram, Fig. 27, measures about 18 feet. The formula for wooden columns shows that this member may be made of a 6″ × 8″ timber, which is amply strong. A 4″ × 8″ timber would probably carry the load, but the length of the strut being more than 45 times the width of the least side, the next larger stock size of timber is adopted, namely, 6 in. × 8 in. The timbers, to facilitate making the connections, should all be of one thickness.

In this case, for instance, the face of all the members in the truss are flush, for all are of the same thickness. The other struts, *KL* and *MN*, will be found, in like manner, to be amply strong, if made of 4″ × 8″ timber. The tie-member is of such a length as to be made up of at least two pieces, spliced at the center. The shear of the wood parallel to the grain determines the strength of the splice, as explained in *Design of Beams*. Bolts may be used to hold the splice together, but should not be depended on to withstand any of the pull on the tie-member.

In estimating the strength of the tie-member at the heel, it may be difficult to obtain a sufficient section of wood at the end of the tie to resist the shear parallel to the grain on the line *ab*. In this instance, the pull on the member is 37,300 pounds. The area of the shearing section on the line *ab* is 8 × 26 = 208 square inches. According to *Design of Beams*, the ultimate shearing strength of yellow pine per square inch, parallel to the grain, is 600 pounds. If a factor of safety as low as 4 is used, the safe strength per square inch will be 150 pounds. Then, 208 × 150 = 31,200 pounds. Now, as previously stated, the stress is 37,300 pounds, and

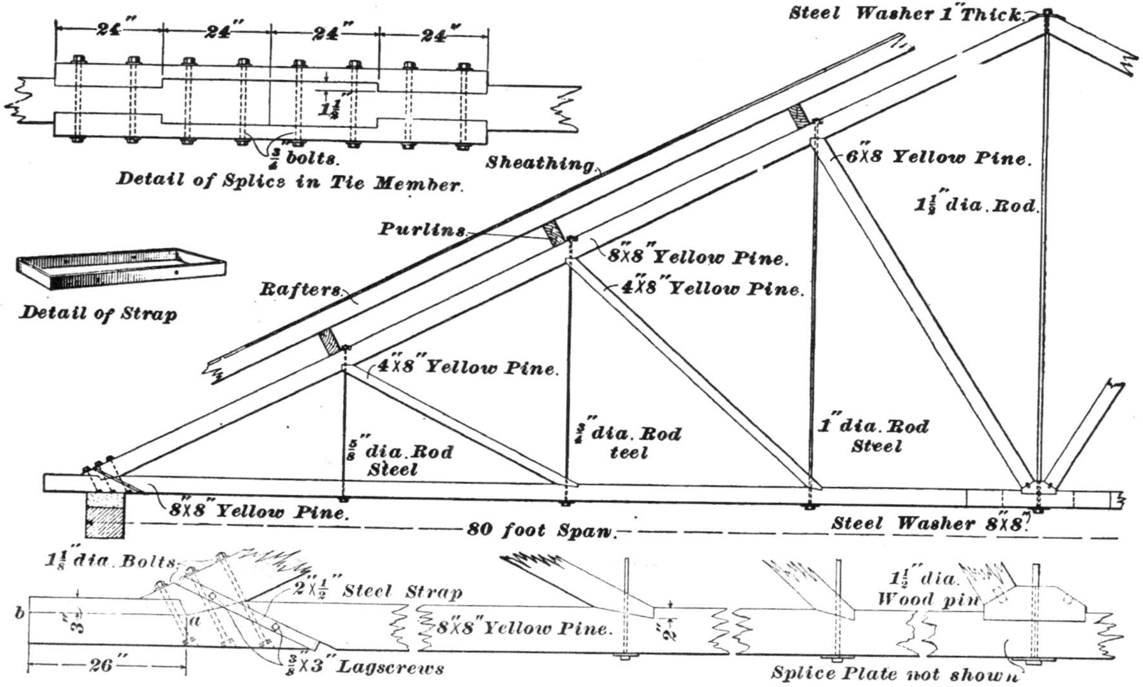
24"
24"
24"
24"
$1\frac{1}{2}$"
$\frac{3}{4}$" bolts.
Detail of Splice in Tie Member.
Detail of Strap
Steel Washer 1" Thick.
Sheathing.
6"X8 Yellow Pine.
$1\frac{1}{8}$" dia. Rod.
Purlins.
8"X8" Yellow Pine.
4"X8" Yellow Pine.
Rafters.
4"X8" Yellow Pine.
$\frac{5}{8}$" dia. Rod
Steel
$\frac{3}{4}$" dia. Rod
teel
1" dia. Rod
Steel
8"X8" Yellow Pine.
80 foot Span.
Steel Washer 8"X8".
$1\frac{1}{8}$" dia. Bolts.
2"X$\frac{1}{2}$" Steel Strap
b
3"
a
26"
$\frac{3}{8}$"X3" Lagscrews
8"X8" Yellow Pine.
2"
$1\frac{1}{2}$" dia.
Wood pin
Splice Plate not shown

FIG. 54

the remaining stress of 6,100 pounds will have to be taken up by securely bolting the joint and by strapping it, as shown in Fig. 54. It is seldom found that sufficient shear can be obtained in the wood to resist the stress at this point, it being always necessary to depend more or less on the bolts. In a large truss, like that under consideration, the best practice demands that the steel strap be so proportioned as to be capable of resisting all the stress not borne by the shear of the wood on the line *a b*, no reliance whatever being placed on the bolts. The lagscrews simply retain the straps in position. One notch in the tie, to receive the rafter, is

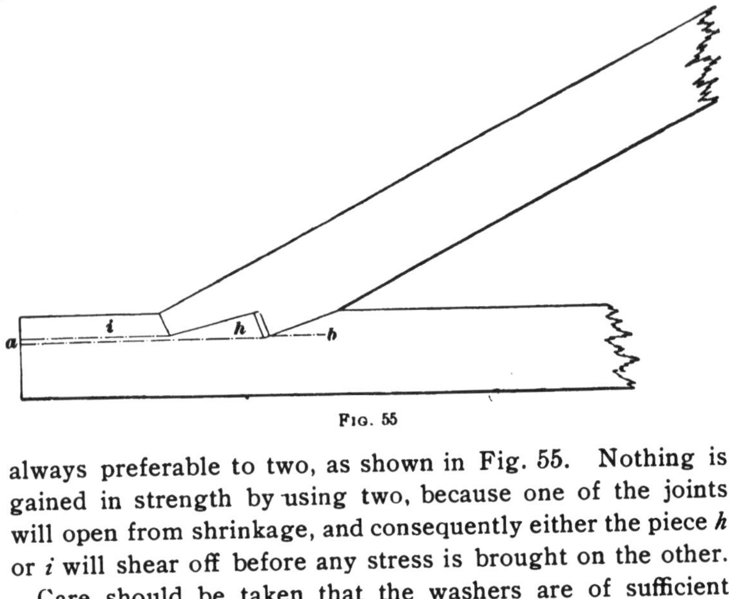

Fig. 55

always preferable to two, as shown in Fig. 55. Nothing is gained in strength by using two, because one of the joints will open from shrinkage, and consequently either the piece *h* or *i* will shear off before any stress is brought on the other.

Care should be taken that the washers are of sufficient area to prevent them from cutting into the wood. In *Design of Beams* will be found a table giving the allowable compression perpendicular to the grain that is used in determining the sizes of the washers.

The tension member $\frac{5}{8}$ inch in diameter, Fig. 54, is not needed to resist any stress, but is useful in the truss to hold the strut in position and prevent trouble from any shrinkage that may occur. No further details need be given in the design of wooden roof trusses, good judgment and foresight on the part of the designer securing, with the principles and practice already laid down, the desired result, namely, maximum strength in the structure with a minimum amount of material.

THE DESIGN OF A COMPOSITE PIN-CONNECTED ROOF TRUSS

34. One of the forms most generally used for a composite pin-connected truss is the Fink truss, previously described. A truss of this pattern is usually made of wooden rafter members, structural-steel struts, and steel tension members. Where the wooden rafter member is oiled, or otherwise finished, and where the steel members are tastily painted, these trusses make a good appearance from the interior of the room or building that they span, and are adapted for supporting the covering, or roof, for such buildings as mess halls of barracks or asylums, armory drill halls, railroad stations, etc.

In order to illustrate the method of designing the members and joints in a truss of this character, we will now consider the design of the truss, shown in Fig. 34, the stresses in which were determined by the diagrams in Figs. 35 and 36, and tabulated in Art. **25.**

35. The **rafter member** has the greatest compressive stress on it between the points *A B* and *B C*, Fig. 34. This stress is represented in the table of stresses by the stress of 80,500 pounds in the member *B K*.

In addition to this compressive stress, there is a bending stress in the rafter due to the construction of the roof, which is composed of sheathing laid on joists spaced about 14 inches center to center along each rafter. The wind and vertical loads are, by this construction, transmitted to the panel points by the transverse strength of the rafter between these points; it will therefore be necessary, before proportioning the rafter member, to calculate the stresses due to the bending moment produced by the dead and wind loads.

The wind load acts normally to the rafter, but the dead or vertical load does not. If great accuracy were required, it might be advisable to resolve the vertical load so as to determine its component normal to the roof, and add this amount to the wind load to obtain the entire load normal to the rafter; such exactness, however, will not be necessary here, and the vertical load and wind load will be added together directly and considered as a uniformly distributed load on the rafter member between the panel points.

The sum of the wind load and the dead load for a panel is 5,625 + 4,500 = 10,125 pounds; therefore, the bending moment, according to the formula $M = \frac{WL}{8}$, is

$$M = \frac{10{,}125 \times 11.188}{8} = 14{,}160 \text{ foot-pounds}$$

or 14,160 × 12 = 169,920 inch-pounds.

Assuming that the rafter member is made of yellow pine, it will be seen by referring to *Design of Beams*, that the modulus of rupture is 7,000; hence, if a factor of safety of 6 is adopted, the safe working transverse stress in the material will be 7,000 ÷ 6 = 1,167 pounds. Since the required section modulus K may be obtained by dividing the bending moment M, in inch-pounds, by the safe unit stress S of the material, as expressed by the formula $K = \frac{M}{S}$, we have

$$K = \frac{169{,}920}{1{,}167} = 146$$

the section modulus required to resist the transverse stress on the rafter member.

Assume that a depth of 14 inches is adopted for the rafter member; then, by transposing the formula $K = \frac{bd^2}{6}$ to $b = \frac{K \times 6}{d^2}$, the breadth, or width, of the rafter member required to resist the transverse stress, by substituting the values of K and d, is

$$b = \frac{146 \times 6}{14^2} = 4\tfrac{1}{2} \text{ inches}$$

Thus, it is seen that a section of $4\frac{1}{2}'' \times 14''$ yellow pine is sufficient to take care of the transverse stress on the rafter member. But, in addition to this transverse stress, there is a direct compressive stress of 80,500 pounds, as shown by the stress diagram, which must be provided for in the same member, by adding material in the direction of its width.

From *Design of Beams*, the ultimate compressive strength of yellow pine parallel to the grain, is found to be 5,000 pounds per square inch, and as a factor of safety of 5 is used, the safe compressive strength will be $5{,}000 \div 5 = 1{,}000$ pounds per square inch.

Since the distance from joint to joint is not great, being only about 11 feet, the rafter member between the joints need not be considered as a long column, but it may be designed to resist direct compression and the full allowable compressive strength of the material parallel to the grain may be used in calculating the cross-section required. Therefore, $80{,}500 \div 1{,}000 = 80.5$ square inches is the sectional area of the material that must be added to the rafter member to resist the compressive stress. The depth of the rafter being 14 inches, and the sectional area required 80.5 square inches, the width to add to the rafter will be $80.5 \div 14 = 5\frac{3}{4}$ inches.

By combining, the total width of the timber required to resist the two stresses is $4\frac{1}{2} + 5\frac{3}{4} = 10\frac{1}{4}$, say 10 inches. The rafter member will therefore be made of one piece of $10'' \times 14''$ yellow-pine timber, which will extend from the heel to the apex of the truss.

36. The **tension rods** should be made of soft steel with eyes formed on the ends and provided with turnbuckles where required; their sectional area should be calculated to safely carry the loads indicated in the table of stresses.

It is well to make all the tension bars in the lower chord in pairs; the stresses on ZK, ZM, and ZQ, Fig. 34, will thus each be taken care of by two bars, one on each side of the rafter member and strut.

Since the stress in the pair of members ZK is 78,250 pounds, each of the two rods composing this part of the truss is proportioned so as to sustain one-half of 78,250, or 39,125 pounds. We will assume that a quality of steel is used whose ultimate tensile strength is 56,000 pounds per square inch, and, owing to the reliability of the material, a factor of safety of 4 is sufficient in the steel tension members; the allowable tensile strength of the soft steel is, therefore, $56,000 \div 4 = 14,000$ pounds per square inch, and the required sectional area of one rod is $39,125 \div 14,000 = 3$ square inches. If square rods are used, as they will be in this case, a $1\frac{3}{4}'' \times 1\frac{3}{4}''$ rod will be found adequate, as it has a sectional area of $1.75 \times 1.75 = 3.06$ square inches.

The size of the other tension rods may be found in a similar manner, but as the stresses on LM and NO, Fig. 34, are light, these members may be made of a single tension bar; these bars and also QZ must be provided with turnbuckles, which will be required to tighten the whole structure and take care of any slack in the truss due to inaccuracy in construction. In designing the members provided with turnbuckles, care should be taken to see that the rods are upset on the ends, so as to realize a sectional area at the root of the screw thread equal to the required sectional area of the rod.

37. The **steel struts** may be proportioned by the formula for calculating the strength of structural-steel columns with one end hinged and one end fixed (see formula given in *Design of Columns*); all the values in the formula are known except the square of the radius of gyration, which may be obtained from the moment of inertia. Assume some convenient section that will be thought to have the required resistance. In this case, the section shown in Fig. 56 is assumed, it being convenient for making the various connections to the tension rods and wooden rafter member.

In order to calculate the least value of R^2, it will be necessary to calculate the least moment of inertia of the section, which will be done in accordance with the principles given in *Design of Columns*.

The properties of this section are given in the tables in *Design of Columns* and the moment of inertia of one of these channels, with respect to the axis *a b*, is 13. Since the neutral axis *a b* of the section passes through the center of gravity of the two channels, the moment of inertia of the section, with respect to this axis, is equal to the sum of the moment of inertia of the channels, with respect to the same axis, that is, to $13 \times 2 = 26$.

The moment of inertia I' of the channel, with respect to an axis through its center of gravity parallel to *c d*, is .7, the area of the section of the channel is 2.38 square inches, and the distance of its center of gravity from the back of the web is .52 inch. The distance of the axis through the center of gravity of the channel, from the neutral axis *c d* of the column section is, therefore, $5\frac{1}{2} \div 2 + .52 = 3.27$ inches. By applying the formula given in *Design of Columns*, the moment of inertia of one of the channels, with respect to the axis *c d*, is

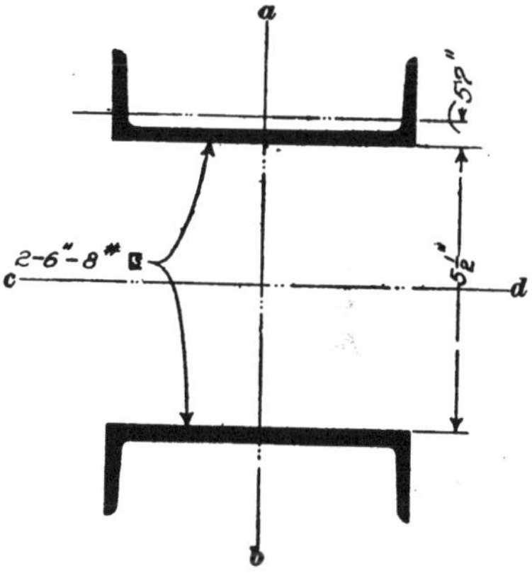

FIG. 56

$$I' = .7 + 2.38 \times 3.27^2 = 26.15$$

For the entire section, the moment of inertia, with respect to *c d*, is equal to the sum of the moments of its parts, that is, to $26.15 \times 2 = 52.3$.

From these calculations, it is evident that the least moment of inertia is on the axis *a b*, and is equal to 26. Then, by substituting the values of the least moment of inertia and the total area of the section in the formula in *Design of Columns*, the square of the least radius of gyration is

$$R^2 = \frac{26}{2 \times 2.38} = 5.46$$

According to *Design of Beams*, the ultimate compressive strength of structural steel is 64,000 pounds per square inch; the length of the longest column in the truss is 8 feet, or

96 inches. By substituting in the formula in *Design of Columns*, the ultimate compressive resistance of the section is

$$S = \frac{64{,}000}{1 + \left(\frac{64{,}000 \times 96^2}{10 \times \frac{9}{4} \times 29{,}000{,}000 \times 5.46}\right)} = 54{,}890 \text{ lb. per sq. in.}$$

Since the area of the section is 4.76 square inches, the ultimate resistance of the column will be $54{,}890 \times 4.76 = 261{,}276$ pounds; if a factor of safety of 4 is used in this column, its safe resistance will be $261{,}276 \div 4 = 65{,}319$ pounds.

Since the stress in this long strut, or column, is only 19,250 pounds, it can readily be seen that it has several times the required strength. It is, however, deemed advisable to use this size of column, as the detailing where it joins the rafter member and also the pin connections at the lower end demand that channels of this size be used.

38. On account of the facilities in making the connections, and because, in a case of this kind, it is generally advisable to adopt the same rolled sections throughout, wherever possible, the short struts should be made of the same rolled sections as the long strut; this method saves labor in the shop and facilitates assembling and erection in the field.

39. The size of the pins is yet to be determined. This was so thoroughly treated in Arts. **30** to **32** that no further explanation will be required here. It is sufficient to say that on thorough examination of the several pinned joints, it was decided that a 3-inch pin would be sufficiently strong to resist any bending, shearing, or bearing stresses that would be applied to them.

The correct design of the castings at the heel and apex of the truss is more a matter of experience and good judgment than of calculation.

40. The truss details shown in Fig. 57 should be carefully studied. They have been designed according to the preceding diagrams and calculations. It is well to observe how all the connections are made and especially the details of the pin connections and the castings at the apex and heel of the truss.

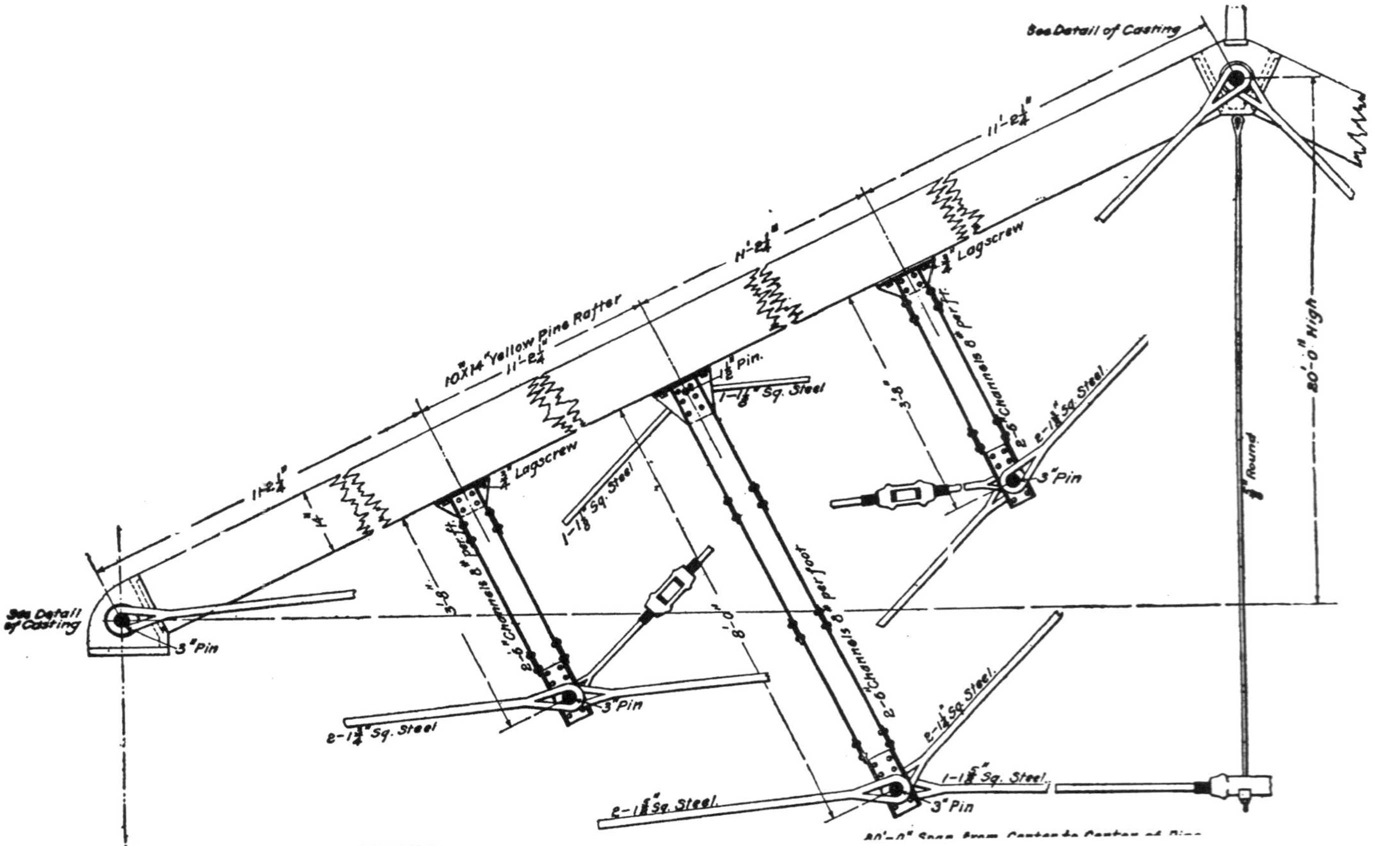
See Detail of Casting
11'-2¼"
11'-2¼"
10"x14" Yellow Pine Rafter
11'-2¼"
11'-2¼"
14"
¾" Lagscrew
¾" Lagscrew
1½" Pin.
1-1⅛" Sq. Steel
1-1⅛" Sq. Steel
2-6" Channels 8# per ft.
2-6" Channels 8# per ft.
2-6" Channels 8# per foot
3'-8"
3'-8"
8'-0"
2-1⅛" Sq. Steel
2-1¼" Sq. Steel
2-1¼" Sq. Steel
2-1⅝" Sq. Steel.
1-1⅝" Sq. Steel.
3" Pin
3" Pin
3" Pin
3" Pin
⅞" Round
80'-0" High
See Detail of Casting

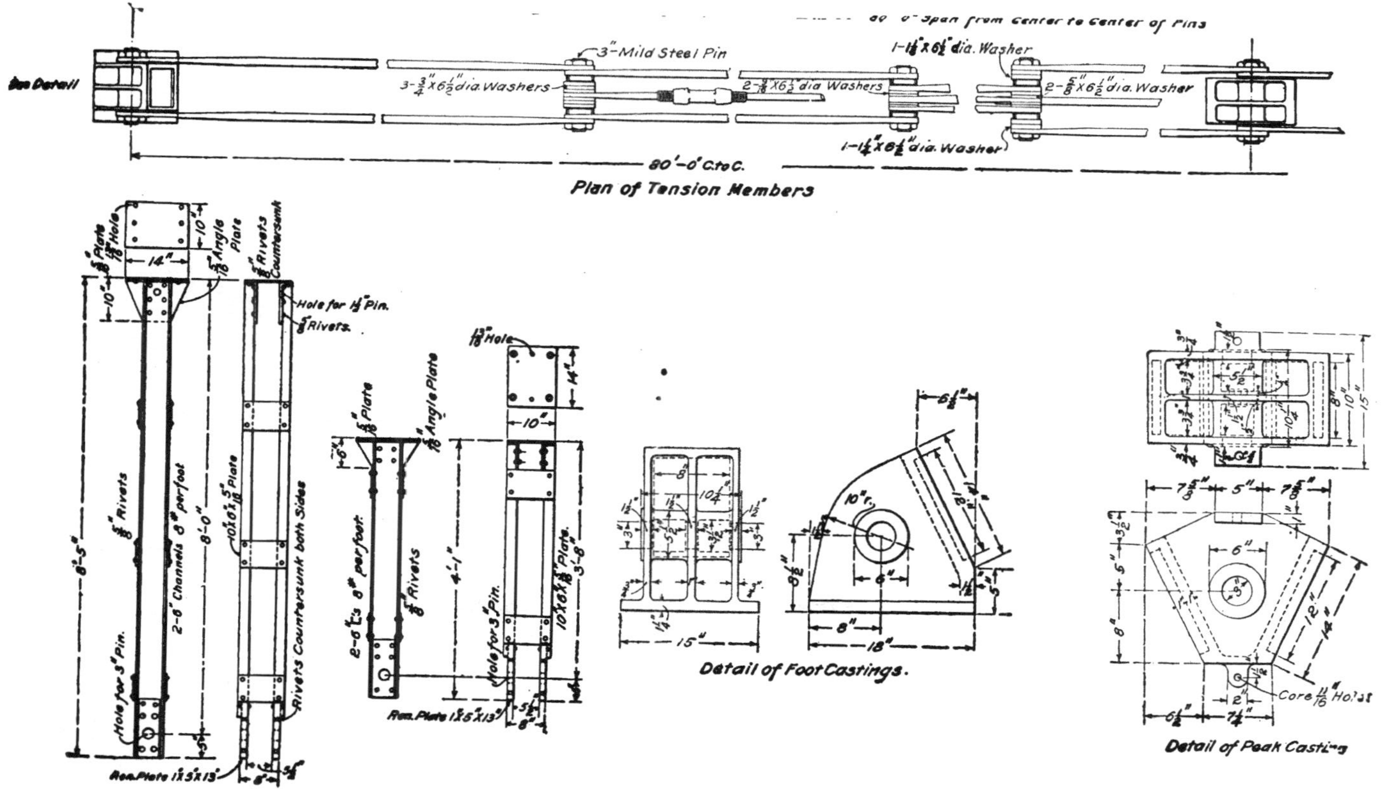
80'-0" Span from Center to Center of Pins
See Detail
3"-Mild Steel Pin
3-3/4"X6 1/2" dia. Washers
2-9/16"X6 1/2" dia Washers
1-1 1/8"X6 1/2" dia. Washer
2-5/8"X6 1/2" dia. Washer
1-1 1/4"X6 1/2" dia. Washer
80'-0" C. to C.
Plan of Tension Members
Hole for 1 1/2" Pin.
5/8" Rivets.
Rivets Countersunk both Sides
2-6" Channels 8# per foot
5/8" Rivets
Hole for 3" Pin.
2-6" L's 8# per foot.
Hole for 3" Pin.
Detail of Foot Castings.
Detail of Peak Casting

FIG. 87

The light rod at the center of the truss has no stress on it, but is used simply to support the lower central tie-member and prevent it from sagging.

In designing compression members made up of two channels tied together with plates, as shown in Fig. 57, the ties should not be farther apart than sixteen times the width of the flange; for example, in this case, the width of the flange of the channels composing the column is about $1\frac{3}{4}$ inches, and sixteen times this width is 28 inches; hence, the distance between the two pieces or ties should not, in this column, be over 28 inches. An inch one way or the other, however, would make very little difference.

THE DESIGN OF A STRUCTURAL-STEEL ROOF TRUSS

41. The material most generally used in the construction of roof trusses for the support of the roof covering of modern buildings is structural steel. The rolled sections

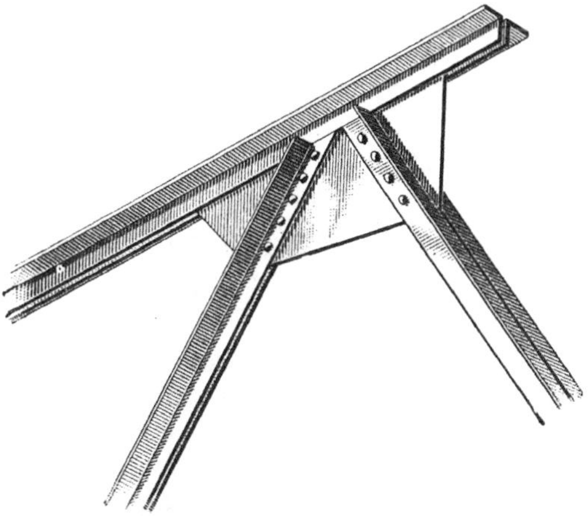

FIG. 58

chiefly used in their construction are angles and plates, though any of the other steel sections may be adapted to special cases.

When a structural-steel roof truss is made of angles and plates, the angles are usually connected in pairs with the plate between them, as shown in Fig. 58. When this construction is adopted, the joints and connections of the several members may be made quite conveniently.

Assume that it is desired to construct the Fink roof truss, previously described and designed as a pin-connected truss, of structural steel. The stresses will be the same as given in the table, Art. 25, and the general dimensions as given in Fig. 34.

The frame diagram may be redrawn and the stresses marked on the several members, as shown in Fig. 59.

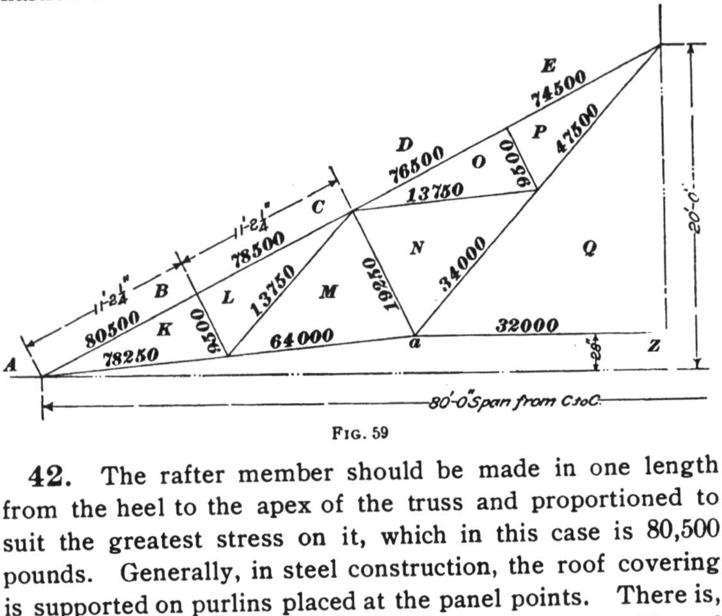

FIG. 59

42. The rafter member should be made in one length from the heel to the apex of the truss and proportioned to suit the greatest stress on it, which in this case is 80,500 pounds. Generally, in steel construction, the roof covering is supported on purlins placed at the panel points. There is, therefore, no bending stress in the rafter, and it is subjected to compression only.

In this case, the rafter is in compression only, and the portion between each panel point will be regarded as a column whose length is equal to the distance from center to center of the joints. Assume the size of a pair of angles that judgment and experience dictate as being adequate to support the stress on the member; then by the short formula for structural-steel columns with fixed ends, determine the strength of the assumed section. If its strength, as found by the formula, is equal to, or slightly in excess of, the strength required to resist the stress in the member, the section may be adopted; provided, of course, that suitable connections can be made to it.

In this case it was decided to assume a section made of two $5'' \times 3\frac{1}{2}'' \times \frac{1}{2}''$ angles, placed back to back with the long legs vertical, and about $\frac{5}{16}$ inch apart. The length of the column is 11 feet $2\frac{1}{4}$ inches, say 134 inches; and from *Design of Columns*, the radius of gyration of a section composed of two $5'' \times 3\frac{1}{2}'' \times \frac{3}{8}''$ angles, with the long legs back to back and $\frac{1}{2}$ inch apart, is found to be 1.51, a value sufficiently exact for our purpose. Assume that the angles are of medium steel. Now, 134 is more than fifty times 1.51. The formula to use from the *Design of Columns* is, therefore, $S_{\prime} = 15{,}000 - 57\frac{L}{R}$. Substituting the value for $\frac{L}{R}$, it is

$$15{,}000 - 57\frac{134}{1.51} = 15{,}000 - 5{,}058 = 9{,}942 \text{ pounds per square inch of section}$$

The area of a $5'' \times 3\frac{1}{2}'' \times \frac{1}{2}''$ angle, according to the table given in *Design of Columns*, is 4 square inches; the area of the entire section is, therefore, $4 \times 2 = 8$ square inches, and the allowable strength of the member $9{,}942 \times 8 = 79{,}536$ pounds. Since the stress on the member is only 80,500 pounds, it is evident that the assumed section will approximately fulfil the requirements, and may be used for the rafter member.

43. The main strut, or the member MN, is the next compression member of any considerable size; its length is 8 feet 7 inches, or 103 inches; it is assumed that a section composed of two $3'' \times 2\frac{1}{2}'' \times \frac{1}{4}''$ angles, placed back to back at a distance apart of $\frac{5}{16}$ inch, will suit this position in the truss. If the two long legs are placed back to back, the least radius of gyration is .95 and is independent of the distance apart that the angles are placed. This value of R is less than $\frac{1}{150} \times 103$. Using the same formula as before,

$$S_l = 15{,}000 - 57\frac{103}{.95} = 15{,}000 - 6{,}180 = 8{,}820 \text{ pounds per square inch of area}$$

Since the area of the two $3'' \times 2\frac{1}{2}'' \times \frac{1}{4}''$ angles is 2.88 square inches, the allowable strength of the section will be 8,820 $\times$ 2.88 = 25,402 pounds. Since the stress in the member is only 19,250 pounds, it is evident that this section will be sufficiently strong.

44. The stress on the short struts KL and OP is so small that any calculation of their required section would only result in obtaining a section composed of such small shapes that their use would not be practical in a truss of this size and character.

Here it is well to call attention to the practical principle that in selecting the section of a member where the stresses are light, care must be taken to choose such a section as will best fulfil the requirements of the construction, disregarding the fact that it will be stronger than is actually required to sustain the load. For example, it is considered good practice to make the leg of the angle held by a rivet not less in width than three times the diameter of the rivet; accordingly, a $\frac{3}{4}$-inch rivet should not be used in an angle leg whose width is less than three times $\frac{3}{4}$ inches, or $2\frac{1}{4}$ inches.

In the case of the truss under consideration, it was deemed advisable to use two $2\frac{1}{2}'' \times 2'' \times \frac{1}{4}''$ angles for each of the struts KL and OP.

45. The tension members in the truss are to be composed of rolled sections similar to those used in the struts; that is, two angles back to back and far enough apart to allow the $\frac{5}{16}$-inch **gusset plate,** forming the means of connection at the joints, to slip between them.

So far as practicable, tension members, in common with the struts, are made up of one continuous section; thus, irrespective of the fact that the stress in ZK is greater than in ZM, these members should both be made of one continuous pair of angles; by so doing, the labor is minimized, thus effecting a saving that would more than offset the cost of the superfluous material in the member ZM, besides producing a much more pleasing appearance.

In the truss under consideration, ZK and ZM will be made of the same pair of angles; QN and QP will also be composed of a single pair of angles, care being taken to proportion the section to withstand the greater stress.

The stress in the member ZK is 78,250 pounds; the allowable tensile strength of structural steel, if the ultimate stress is taken at 60,000 pounds, and a factor of safety of 4 is used, will be $60{,}000 \div 4 = 15{,}000$ pounds per square inch. Then $78{,}250 \div 15{,}000 = 5.21$, the sectional area in square inches that will be required in the pair of angles forming the member ZK after deducting the sectional area cut out by one rivet hole in each angle.

From *Design of Columns*, a $3'' \times 3'' \times \frac{9}{16}''$ angle is seen to have a sectional area of 3.06 square inches; two angles will then have a sectional area of twice 3.06, or 6.12 square inches. Since the thickness of these angles is $\frac{9}{16}$ inch and a $\frac{3}{4}$-inch rivet is to be used, the area of the section to be deducted for one rivet hole is $.875 \times .5625 = .49$ square inch, and the net sectional area in the two angles is therefore

$$6.12 - 2 \times .49 = 5.14 \text{ square inches}$$

This is slightly under the sectional area demanded by the calculation, but, as a low ultimate tensile strength was assumed, it will be safe to use this section for the members ZK and ZM.

The other tension members in the truss may be proportioned in the same manner, care being taken to deduct from the section area of the member the sectional area removed for rivet holes; also, that the rolled sections adopted will satisfy the practical demands of the construction.

46. The number of rivets required at the several joints in the truss should be carefully calculated. Since the method of calculation is the same for all the joints, only one will

plate, which, according to Table III, is 3,656 pounds, with an allowable stress of 15,000 pounds. The value of four rivets will then be $3{,}656 \times 4 = 14{,}624$ pounds, a little less than the amount to be taken care of, but the difference is so small that it may be disregarded and for the sake of symmetry the same number of rivets will be used in connecting the splice plate to the member *MZ*. The number of rivets required to connect the member *ZQ* to the gusset plate may be readily obtained.

47. Fig. 61 shows the usual shop drawing of the truss just designed. Particular attention should be paid to the details of the connections. **Separators** should be placed in both the tension and the compression members; they are placed in the tension members to prevent the angles from striking against each other when the trusses are subjected to vibrations, and also to join the two angles of a member, so that the work will arrive at a point of erection in a convenient form, ready to be put together. The separators are placed in the compression members so as to insure against any tendency to bend them apart, so that they will act in unison. The spacing of these separators is more a matter of judgment on the part of the designer than anything else, though any spacing over eight times the least dimension of the member is not to be recommended. Separators should always be placed near the end of a pair of angles to be connected to a gusset plate, as it will hold them the right distance apart in shipment and facilitate erection in the field. The $1\frac{3}{4}'' \times 1\frac{3}{4}'' \times \frac{1}{4}''$ angle, joining the apex of the truss and its lower chord, is required only to support the chord angles and prevent them from sagging. When there is considerable stress in this member, sagging is unlikely to occur, but it is the usual practice and a good one, to introduce some support for this long and usually light member. It will be noticed that one size of rivets is used throughout this truss wherever practicable; this is a point in economical construction that should always be observed. The ends of the angles and members, when practicable, should be cut off square, and the gusset plates should be designed so that they may be formed with as few cuts as possible, and unless it is desired to make the truss somewhat fanciful, these cuts should never be other than straight lines.

In designing roof trusses, and, in fact, all structural-steel constructions, care should be taken to see that the several sections into which each may be divided are not so large that they cannot be shipped by railroad or other transportation at hand; this is important and should be carefully considered.

GENERAL NOTES REGARDING THE DESIGN OF A ROOF TRUSS

48. Lateral Bracing.—Trusses forming the principal support of a roof, if of any considerable size, should be braced together in the planes of the rafters so as to secure them against any tendency of the wind, when blowing in a direction perpendicular to the gable ends, to produce lateral movement. If the roof sheathing is laid close and is well nailed, it will sufficiently stiffen trusses of moderate span. The heels of trusses are sometimes fastened securely to the walls, especially in those buildings where the wind is liable to get under the roof. When so secured there is a tendency for the wind to reverse the stresses in the members of a roof provided with a light covering, and this reversal should be taken care of in the design of the truss.

49. Factors of Safety.—Since due allowance must be made for unforeseen and unknown defects of material and workmanship, and for unknown stresses that are liable to occur, it is necessary to proportion the several parts of a structure so that they will be able to resist, without failure, much larger forces than those obtained from the stress diagram.

In roof trusses, however, the stresses can be calculated with more certainty than in the case of a bridge or a machine, and the application of these stresses is more steady in its nature, and, therefore, not so severe on the material. For this reason, it is permissible to allow unit stresses, in the design of roof trusses, some 50 per cent. in excess of those considered allowable in first-class bridges.

50. Tension Members.—The strength of a tension member is that of the smallest cross-section. It is therefore good practice to *upset*, or enlarge, the ends of long bolts or tension rods with screw ends, so that the cross-section at the root of the thread will be at least equal to the sectional area of the main part of the rod.

The central axis of the cross-section of ties and struts should coincide with the line of action of the thrust or pull, as otherwise the piece will be greatly weakened and dangerous bending moments will be developed in the structure. To calculate the net sectional area required in any tension member, the force or load on it should be divided by the safe working tensile strength of the material, and allowance must be made in the area of the cross-section for the cutting of bolt and rivet holes.

51. Compression members whose lengths do not exceed six times the least dimensions may be proportioned by the method followed for the tie as just explained, but, when the length of the truss is increased, there is a tendency to yield sidewise when compressed, and the sectional area must be increased or the unit stress diminished. Hence, the column formulas given in *Design of Columns* must be used. Pieces subjected to alternate compression and tension should have a materially larger section than would be required for either stress alone. Cast iron is seldom used in the best work for anything but short compression pieces, packing blocks, and pedestals.

52. Members in Trusses Subjected to Transverse Stresses.—In determining the resisting moment of a member subjected to transverse stresses, or in calculating the section required at the point of maximum bending moment, due allowance must be made for portions cut away on the tension side in attaching fastenings, or in making connections; similar allowance must also be made on the compression side, unless the holes are completely filled by the rivets, in which case no deduction need be made from the sectional area of the member.

53. Members in Trusses Subjected to Both Transverse and Direct Stresses.—The rafter members in a roof truss, likewise members in other structures, are often called on to resist both a bending and a direct stress. Such pieces must first be designed to safely resist the bending moment, and then their transverse dimensions must be increased so that the added material will have sectional area sufficient to resist the direct pull or thrust. Should the direct force be a compressive stress, it will be well, after the member is designed, in order to check up its strength, to test by the proper column formula.

54. Pins and Eyes.—In proportioning the pins and eyes of tension bars, the diameter of the pin should be from three-fourths to four-fifths of the width of the bar in flats, and one and one-fourth times the diameter of the bar in rounds. The sectional area of the metal around the eye should be 50 per cent. in excess of that of the rod or bar. When flat bars are used, their thickness should be not less than one-fourth of their width; this will secure a good bearing surface on the pin. The size of a pin is usually decided by the bending moment on it, consequently the assembled pieces on a pin should be packed close together, and opposing members should be brought as nearly in line with one another as possible.

55. Details.—The designer should carefully examine each joint and connection in the structure, and consider such practical points as means of shipment, erection in the field, etc. Care should be taken in designing all connections, to so place the rivets and bolts as to realize their full strength, and at the same time not cut away too much of the material of the members connected. All members and joints should be examined for tension, compression, shearing, and bending, and proportioned accordingly. The strength of the joints and connections between the members of a structure are of as much importance as the strength of the members themselves; the strength of any structure depends on the strength of its weakest point, and its failure at a joint or connection is as fatal as the failure of any of its members.

The designer will find the handbooks issued by the various steel mills of great value in the prosecution of his work. They contain many useful tables giving the properties of rolled sections, with information as to their use and application.

MILL DESIGN

Serial 1097-2 Edition 1

SITE AND ARRANGEMENT

PRELIMINARY CONSIDERATIONS

INTRODUCTION

1. The requirements of the modern factory building are many, and demand the careful attention of the architect in their planning and construction. There are probably more rules and regulations imposed by the state and local governments and by the Insurance Underwriters, regulating the construction of this class of buildings, than for buildings of any other character.

The laws imposed by the governments under whose jurisdiction the building is to be erected, are framed manifestly for the protection of the health and safety of the occupants of the building, and so as not to jeopardize their lives in case of fire or panic, or the lives of those engaged in the attempt to save the structure and prevent damage to the adjoining property.

The Underwriters, or the Association of Insurance Companies, have compiled numerous rules and regulations of which the architect planning the building must take cognizance if he desires to secure a reasonable rate of insurance on the building and its contents for the owner. Not only do these rules and regulations deal with the structural design of the building, but they consider the apparatus for protection in case of fire, and such installations as the electric

COPYRIGHTED BY INTERNATIONAL TEXTBOOK COMPANY. ALL RIGHTS RESERVED wiring. The architect must be familiar with all these requirements in order to intelligently and practically design industrial plants.

There are, also, many factors essential to the utilitarian and economic operation of the building entering into the design of the modern factory, to which the architect must devote careful study. Among the most important of these are the economic receiving, shipping, elevation, and transportation of merchandise; the proper and adequate lighting of the building; the location and planning of the power plant for the building, together with the engineering problems of construction, which include the design of the floor, columns, and walls for the loads to which they are subjected.

CLASSIFICATION OF FACTORY BUILDINGS

2. Classified according to their construction, factory buildings may be divided into three types, which, for convenience, may be designated as *first-*, *second-*, and *third-class buildings.* A similar division to this is also frequently made by the state or municipal laws for the regulation of the construction of factory buildings.

3. First-Class Buildings.—Buildings of the first class are those in which the walls, floors, columns, girders, beams, partitions, and roofs are of stone, brick, terra cotta, concrete, steel, iron, and such other fireproof materials as have been proven to be efficient. Buildings of this class may be considered as constituting an entirely fireproof building, which means that while the contents of the building may burn, the building itself will remain intact, unless subjected to the severe action of a prolonged conflagration.

4. Second-Class Buildings.—Buildings of the second class are considered to include what is known as slow-burning, or the typical factory-construction, type, in which all posts or girders must be of heavy and massive timber, and the floor construction at least 3 inches in thickness, and of solid planking. In buildings of the second class, while it is permissible to use combustible materials, they must be of such sizes and of such slow combustion that the security of the building will be insured for a reasonable time after the conflagration has commenced. It is usual, therefore, in this class of building, to limit the size of the wooden posts to not less than 8 inches square, though their strength may be greatly in excess of the load they are required to support, and girders and beams are used whose least dimension is 6 inches or more. In buildings of this character, it is frequently necessary to use steel beams and columns in order to obtain the strength for the great floor loads to which these members are liable to be subjected. When such steel or iron columns are used, however, they must be fireproofed, because even though made of incombustible material, they would not have the same endurance in a fire as have heavy wooden girders or posts, and their failure would precipitate the fall of the floor. Wooden girders and posts, even when charred part way through, have still sufficient strength for the support of the load for which they were designed.

5. Third-Class Buildings.—Buildings of the third class are not particularly recommended for the construction of factory buildings, for the floors of these may be of the ordinary joist and finished floor construction. Such buildings are readily ignitible and burn rapidly, not only because the timber work in them is light, but because of the numerous air spaces that exist between joists and in the furring of the walls. No building with air space surrounded by combustible materials can be considered as slow burning.

FACTORY PLANNING

6. Considerations in Planning.—The outline of the building is determined by the site, and owing to irregularities in the site generally purchased for manufactory purposes, it is frequently difficult to properly design buildings of this character. The factors that probably influence the design mostly, after the location of the column supports, and consequently the spacing of the windows has been determined upon, are the stairways and elevators.

In many cases, these are the only subdivisions of the main-floor plans, and in order to comply with the rules and regulations of the local or state governments, and of the Underwriters' Association, they demand primary consideration. The stairways must as well be easy of access, while the elevators must be conveniently located for the delivery and receipt of goods from the first floor of the building.

Another factor that is likely to enter into consideration of the design is the toilet rooms, which must be placed against an outside wall, and convenient to any part of the floor.

ARRANGEMENT OF STAIR TOWERS

7. The Enclosed Stairway.—The important consideration in the planning of factory stairways is to provide a quick, easy, and safe egress for the occupants in case of fire. While it is necessary to have good liberal stairways for communication between the floors, this is not such an important feature in factory design, from the fact that there is little travel of employes between the floors, in a modern factory, as each individual's work is usually apportioned to him and confined to a particular location, and therefore does not require him to be on different floors during the day.

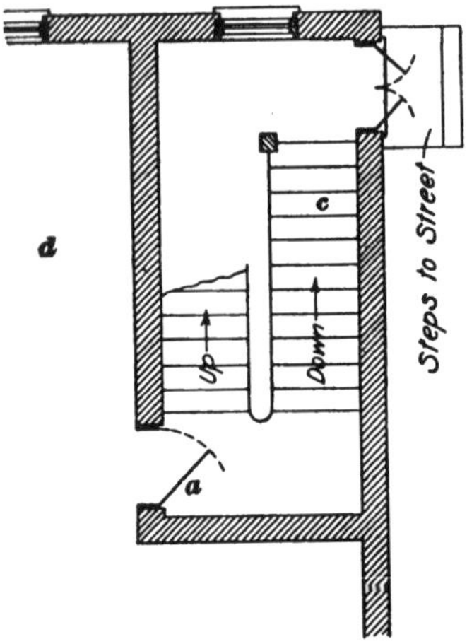

FIG. 1

The common type of factory stairway is that designated in Fig. 1. This shows a brick-enclosed stairway with the doors entering it direct from the factory. Such a stairway enclosure as this should have tin-lined doors as at *a*, which fireproof the opening. Even then the security of a stairway of this character is not certain, from the fact that these doors may be left open, and are open to the stairway during the egress of the occupants. With a severe fire, therefore, on any floor, such a stairway is likely to be filled with smoke to suffocation, and liable to ignition from the door openings. It can therefore only be regarded as a makeshift for a fire-escape, or fire-tower. It is also well, in the design of such a stairway, to observe that the doors

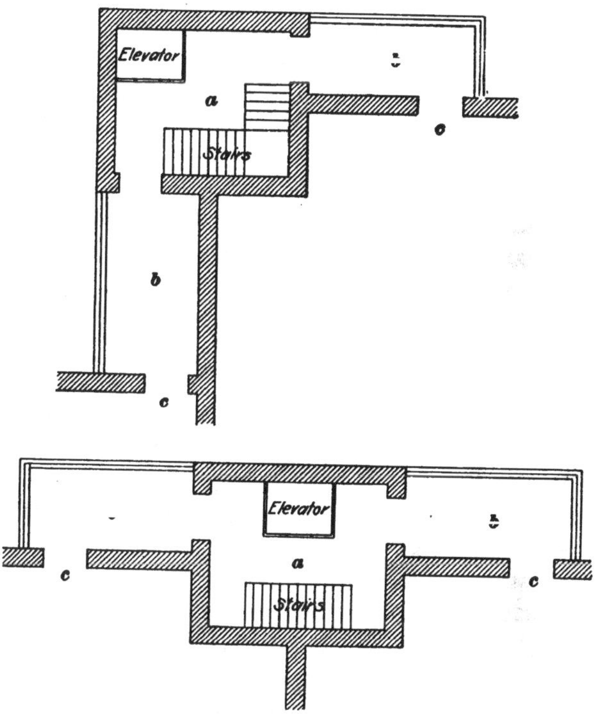

FIG. 2

always open outwards, and not only this, but that they open with the tide of people coming down the stairs. For instance, in the figure the door *a* is opened correctly, but if the flights of steps marked *down* and *up* were transposed, then people coming down the flight *c* would press against the door at *a* and prevent the people in the room *d* from getting out to the stairs and hence to safety.

8. Enclosed Fire-Escape, or Stair Towers.—Various designs for brick-enclosed stairways of factory buildings have been recommended at different times by the insurance companies. Two of these designs are designated in Fig. 2 and show the elevator included, as well as the stairway, in the tower.

A study of these plans will show that there is no direct communication from the building to the stair tower, and that the only way by which the stair tower may be entered is through an open balcony, which communicates with a door in the side walls of the building, at each floor. The brick-enclosed stair tower is shown in the figure at *a*, the open galleries at *b*, and the door of egress from the factory at *c*. By means of this arrangement, the occupants of each floor can, in case of fire, go through the door in the side wall, on to the open balcony, and into the fire-tower, thence down stairs or elevator to the ground floor and safety. Because of the openness of the balcony, which is surrounded with a strong rail, and partially covered by the gallery above, the fire could hardly be sufficiently great on any floor to make it untenantable, and no smoke or flames could communicate with the fire-tower.

9. Vestibule Fire-Tower Stairway.—While the arrangement just described is rational, the fire-tower is of such dimensions as would ordinarily preclude its use in the modern building where ground space is valuable and every inch of surface must be economized.

The best possible design, therefore, for a fire-tower is that indicated in Fig. 3. This construction is known as the **vestibule fire-tower,** and from the plan it is seen that this combines safety and utility in a small amount of space. The opening marked *a* in the plan is always open to the weather, and the floor of the vestibule is usually concreted and graded to a drain that connects with the rain conductor of the roof at *b*. By this arrangement, a well-protected line of travel is obtained between the stairway and the buildings on each floor, the occupants of the building being protected by the parapet wall and iron railing, as indicated at *c, c*. To this vestibule from the factory a tin-lined, or fireproof, door must be provided, so that after the people have left one floor it can be cut off from the vestibule.

In the construction of all fire-towers, their walls must be carried by means of parapet walls at least 3 feet above the roof of the building, and the roof over them must be constructed of fireproof material.

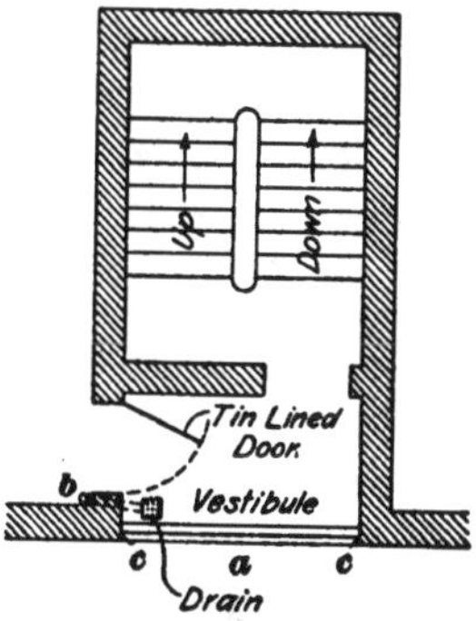

FIG. 3

When it is required, fireproof windows may be used in the walls of brick-enclosed fire-towers. These windows may be constructed of sheet metal and glazed with wire glass. If it is not possible to use such windows, a skylight may be built over the top of the tower, but this skylight must be constructed of sheet metal, or other non-combustible material, and glazed with wire glass.

10. Number of Fire-Towers.—The number of tower fire-escapes required for factory buildings of either the first, second, or third class, may be established according to the number of stories in height of the building, and the floor area, in square feet, for each floor; that is, for buildings of the first class, three or four stories in height, having one tower, the floor area of any floor may be as much as 20,000 square feet, while if the height of the building of the same construction is made twelve stories, the floor area should only be 6,500 square feet.

Where two tower fire-escapes are incorporated in the plan, a building three or four stories in height may contain as many as 25,000 square feet in the area of one floor, but if the building were increased to twelve stories, the floor area of each floor should not exceed 15,000 square feet.

In buildings of the second and third classes, a greater number of tower fire-escapes should be provided, and it is good practice to supply one tower fire-escape in a three-story building of these classes for a floor area not exceeding 10,000 square feet, or two tower fire-escapes should the floor area not exceed 15,000 square feet. In buildings of this construction, of from four to six stories in height, the floor area should not exceed, for one tower fire-escape, from 6,000 to 3,500 square feet of floor area in each floor, while with two tower fire-escapes the maximum floor area is from 12,000 to 8,000 square feet.

11. Location of the Fire-Tower.—Where two tower fire-escapes are used in a building, they must not be located near to each other, the purpose always being to provide a second egress in case of one being cut off by smoke or flame.

In small buildings of considerable height, it is sometimes difficult to so arrange the plan as to provide two stairways at extreme ends or corners of the building. In a case like this, it is frequently necessary to extend balconies along the side of the building, entering the fire-tower at some more distant point.

ELEVATOR SHAFTS

12. Location of Shaft.—In all factories of two or more stories, an elevator is a necessity for the economic transmission of goods from one floor to another. While in some instances the elevator is run through hatch openings in the floor, without being enclosed in brick walls, it is not good practice, for the openings through the floors make possible the rapid communication of flames and smoke in case of fire, even when provided with an automatic closing hatch. Elevators are therefore generally built in an elevator shaft, the walls of which are constructed of good hard brick and made from 12 inches to 18 inches in thickness.

13. Elevator Doors and Openings.—In building elevator shafts, it is necessary to provide door openings at each floor, the openings being protected with tin-lined fireproof doors. These doors may be either folding or sliding doors, it usually being considered best to provide a sliding door that will automatically close when a certain temperature has been reached in the building. It is not always possible, however, to provide sliding doors, from the fact that where the elevator shaft projects out into the room and the door opening is wide, there is no wall space on which to fasten the track. Customarily, in each door opening, there is also provided a heavy stone or cast-iron sill. As in many states the law requires some automatic, or folding, lift gates for elevator openings, it is the practice to project this sill inside of the shaft at least 4 inches, in order to provide a bearing and protection for such gates.

14. Construction of Openings.—The openings in the brick walls of an elevator shaft are constructed ordinarily with rowlock brick arches, and from the fact that the openings are usually wide, and little jamb is left on each side of the opening, it is necessary to build 1-inch round iron rods in the arch above the opening, these tie-rods being furnished with washer plates at each end.

The jambs of all openings in the elevator shafts should be protected with cast-iron fenders made of about ½-inch metal, and so constructed as to return about 4 inches on each face. These jambs are provided with heavy wrought-iron anchors, which are built into the brickwork in the process of construction.

15. Freight Elevators.—Freight elevators are either corner-guided or side-guided, the preference being for the

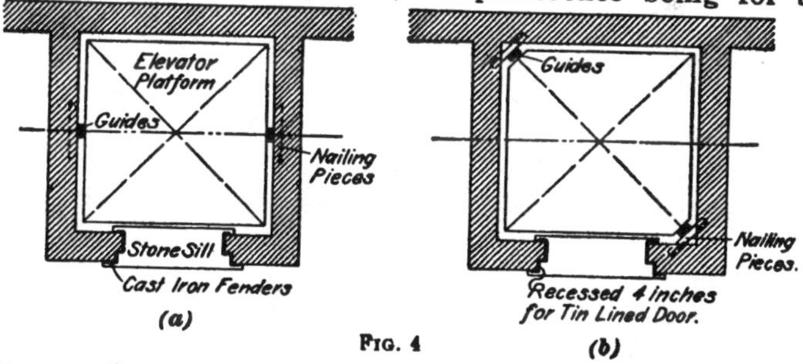

FIG. 4

latter, as they are more easily constructed and adjusted, and during the construction of the elevator shaft it is usual to build into the brickwork blocks of wood about the thickness of a brick and of sufficient length for the attachment of the guides. These are cut wedge-shaped on the ends, so as to hold more firmly in the brickwork. A diagrammatic plan of a side-guided and corner-guided elevator is illustrated in Fig. 4 (*a*) and (*b*), respectively.

Owing to the fact that it is necessary to have considerable hoisting mechanism, at the head of the elevator shaft, the shaft is extended above the roof, sometimes as much as 5 or 6 feet, for the minimum height from the elevator platform at the top floor level to the under side of the beams carrying the mechanism is about 16 feet, and the sheaves carrying the rope and other mechanism at the top of the shaft require several more feet.

16. Elevator-Shaft Windows.—Frequently, elevator shafts are lighted with windows. Where such windows open into the building, they must be constructed with metallic frame and wire glass; but where they open outside of the building it is not necessary to do this. Where elevator shafts are lighted from the top, metallic skylights glazed with heavy glass should be provided. It is not considered such good practice to use wired glass for this glazing, the idea being that in case of fire, as each floor is cut off from the elevator shaft with fireproof doors, vent may be had from the skylight at the top of the shaft when the glass is broken, and some of the municipal and state laws stipulate that the skylight at the top of an elevator shaft shall not be less than two-thirds the area of the shaft.

As there is some mechanism on the bottom of the elevator platform and at the foot of the shaft, it is necessary to sink the shaft at least 3 feet below the basement floor level, provided that the elevator runs to the basement floor.

TOILET ROOMS

17. Location of Toilet Rooms.—In designing factory buildings where a great many people are employed, the question of toilet accommodations is a very necessary consideration.

After the location of the toilet room has been decided on, and it should be placed as centrally as possible to the floor area, the number of closets should be determined. It is usually found sufficient accommodation if one closet is allowed for twenty people. In planning the toilet room, it is essential, and generally required by law, that the room shall be located on the outside wall, so that windows will open directly into it. If the partitions between the compartments only extend part way toward the ceiling, it is not necessary that each compartment of the toilet room containing a closet should have a window. For instance, referring to Fig. 5, which is a typical arrangement of factory

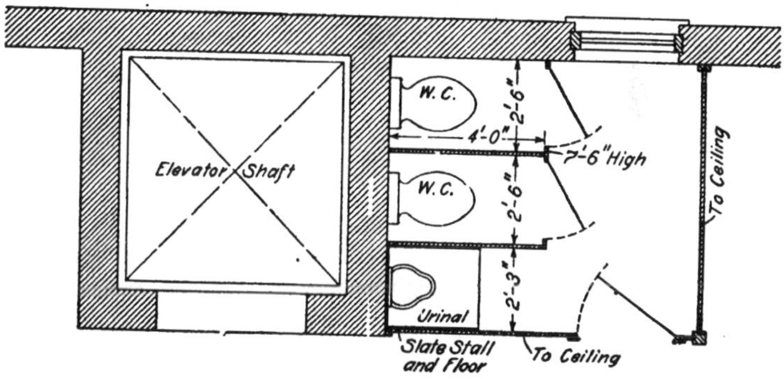

FIG. 5

toilets, it is necessary only to provide the window opening into the outside space surrounding the enclosures around the closets.

18. Material Used for Partitions.—The partitions of both the compartments and the toilet room are, in factory construction, usually built of $1\frac{1}{8}$-inch, yellow-pine, tongued-and-grooved, beaded ceiling, the corners of the partition being braced with 4″ × 4″ stop-chamfered yellow-pine posts rabbeted to receive the ceiling. In no instance should the partitions be constructed with an enclosed space or concealed work. For hygienic reasons, it is always advisable to provide toilet rooms with waterproof floors, and the brick walls that may partially surround the enclosure should be water-proofed for a distance of at least 1 foot from the floor. The waterproofing commonly employed for the floors is asphalt or *asbestolin*, the latter being a composition that is placed directly on the finished floor and forms a permanent covering about ¼ inch in thickness, and of the nature of the best cork linoleum.

In waterproofing the walls, about the only practical method to employ is to give them several coats of an approved waterproof paint.

Fig. 6 illustrates a reasonably cheap and still excellent

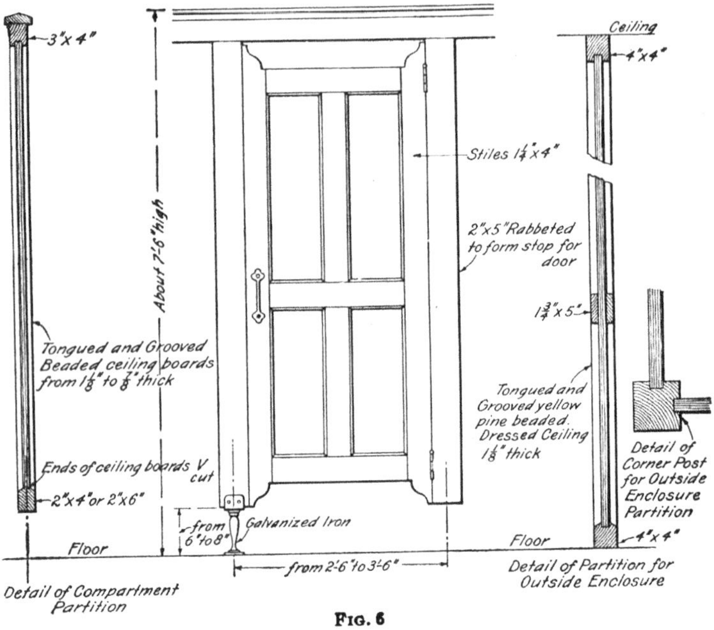

FIG. 6

construction for toilet-room enclosures, the several details of the construction being sufficiently clear without further explanation.

19. Toilet-Room Fixtures.—In selecting fixtures for the toilet rooms of a factory building, only the most serviceable kinds should be used. In standard work, iron-porcelain siphon-jet closets with overhead copper-lined flushing tanks are employed. The flushing is controlled automatically by the action of the seat, which is always raised by being provided with counterweights. Lids to the closets are never used in factory installation, from the fact that they are readily broken and generally out of order.

TYPES OF MILL CONSTRUCTION

GIRDER AND PLANK-ON-EDGE CONSTRUCTION

20. In Fig. 7 is designated an economical type of mill construction, which is much in use. This construction is slow burning in every respect, and is exceedingly simple, withal being substantial and presenting a good appearance on the interior of the building. It will be observed that the column supports of the floor consist of yellow-pine posts, varying from 20 inches square to 8 or 10 inches square, the latter being used for the support of the roof. The drawing shows columns and wall construction suitable for a six- or seven-story factory building, and the size of post indicated in the basement is about the maximum.

The girders consist of $6'' \times 20''$ yellow-pine pieces bolted together with $\frac{3}{4}$-inch bolts, though it is suggested that $\frac{7}{8}$-inch bolts would be preferable. The purpose in bolting up a girder in this way is that the thinner planks are much more readily obtainable, they are more likely to be thoroughly seasoned, and a girder built up in this manner is usually stronger than a solid beam, from the fact that there is less likelihood of hidden defects existing in the timber and much better stock can generally be obtained.

21. Post Caps and Base Plates.—These girders just described are supported on cast-iron post caps, similar to the Goetz-Mitchell construction. These post caps, where they support girders in one direction only, are usually known as two-way caps. If they support girders in two directions—that is, transverse and longitudinal—they are known as four-way caps. These caps are generally cast of $\frac{3}{4}$-inch metal, and the girders bear on them at least 4 inches.

For the basement columns, it is usual to provide a cast-iron base plate, as indicated at *a*. The timber column is sized into the socket of the base plate, and it is best to carry the top edge of the cap well above the floor, so that any moisture from leakage or in washing the floor will not be allowed to penetrate to the wood. Owing to the fact that this base plate must transmit the entire load on the column to the brick, it must be heavily webbed on the sides and corners, as indicated in the plan at *b*.

22. Concrete Footings.—It is usual in designing the foundations for mill buildings to use concrete footings, as indicated in the plan. It will be noticed in this particular instance, and it is the usual practice, that the concrete footings are 12 inches in thickness. When the footings are stepped, as indicated, under the wall of the building, each footing is made about 12 inches in thickness, with a projection of not more than 6 or 7 inches.

23. Floor Construction.—The floor construction of the building consists of 3″ × 6″ yellow-pine pieces, set on edge and spiked together. Such a construction as this is available for spans between girders of from 10 to 15 feet, and does away with all secondary girders, or beams. It also has an advantage in that it presents a neat ceiling beneath when the edges of the planks forming the rough flooring are beveled.

Section of Finish flooring

FIG. 8

On the top of this rough flooring, which is designed for carrying the floor load, and which is so constructed that the joints in the different pieces are broken, a 1-inch or $1\frac{1}{4}$-inch maple flooring is laid. Maple is used for finished floors in factories principally on account of its hardness and the excellent wearing surface that it affords. The maple flooring available in the market runs in lengths of from 3 to

16 feet, but the cost of the floor is greatly increased if a minimum length of 6 or 8 feet is specified. Usually the flooring is tongued and grooved and hollowed on the back, as indicated in Fig. 8. The hollow back prevents the flooring from curling. In a better class of finished flooring, the pieces are end-joined, or provided with a tongue and groove on the end. This prevents the end of the flooring from turning up and interfering with the smoothness of the floor and the operation of trucks over it.

24. Waterproofing and Dust Proofing.—For the purpose of deadening sound, and sometimes for the sake of waterproofing, sheathing paper or felt is inserted between the finished flooring and the rough plank. By the introduction of paper between the maple and the rough flooring, dust and dirt are prevented from falling through the crevices due to the shrinkage of the flooring boards above.

It is usual in finishing a floor around the edge to use about 2-inch quarter-round molding, as indicated at *c*, Fig. 7. This molding is also used around the wooden columns, or posts.

In order to prevent the posts from splintering at the corner, and so that there is less likelihood of the occupants being hurt, a stop-chamfer, or arris, is formed on the corner.

25. Splice Pieces.—There is one feature which must not be overlooked in mill construction, such as occurs in this figure, and that is that since the girders butt against the columns on top of the post caps, usually flush, there is nothing to carry the boards at *d*, therefore yellow-pine pieces or steel angles must be provided, as indicated at *e*. These splice pieces answer two purposes, namely, to form a bearing for the ends of the planks *f*, *f*, *f*, and also to tie the girders rigidly together longitudinally, and thus increase the rigidity of the floor construction.

For a similar reason, it is necessary to form a ledge, either by reducing the size of the pier above, or bv corbeling out, at the window openings, as shown at *g*, for the support of the floor planks at *h*, *h*. On the top of the corbel so formed, usually a 3″ × 8″ yellow-pine piece is securely anchored to the wall, to provide a bearing for the ends of the rough floor planking.

26. Reference to *i* and *j*, Fig. 7, shows that there is very little room between the head of the window and the bottom of the floor construction. By this means, the maximum amount of light near the ceiling is obtained, and, besides, the ventilation is greatly facilitated. This is one of the important features in factory designing, as well as in school-house architecture.

27. Foundation Walls and Piers.—From the section of the wall shown at *k*, *k*, Fig. 7, it will be observed that the entire building is practically supported on heavy piers, and that the 13-inch walls below the window sills are only spandrel fillings. In some instances, 9-inch walls can be used in these places, but it is not considered advisable from the fact that beating rain will readily drive through a 9-inch wall, and, besides, there is hardly sufficient sill for a heavy window frame.

Attention is particularly called to the construction of the window sill at *l*. In the better class of construction, heavy bluestone sills $5\frac{1}{2}$ in. × $7\frac{1}{2}$ in. would be used; but for cheap work, it is customary to use a light 3″ × 5″ bluestone sill.

Where spandrel fillings more than 13 inches in thickness are used, or where the thickness of the wall is much greater than the frame, as indicated at *m* in the basement, beveled bricks on edge are used for forming the sloping wall inside. The purpose of the sloping sill is to prevent the corners from being broken and damaged, and employes from occupying them.

28. Terra-Cotta Window Heads.—In factory construction, the use of terra-cotta window heads is not unusual, and the construction of such a window head is indicated at *n*, Fig. 7. Where terra-cotta window heads are used in this manner, some means of support must be had for the brickwork above the window head, as terra cotta in itself is of little use as an arch, or lintel. It is not uncommon to use angle irons back to back, as indicated on the section of window head *n*. This construction, of course, can only be used where the wall runs parallel with the supporting floor, for if the head of the window receives beams or girders it must necessarily be more strongly and rigidly constructed with heavy channel irons, or I beams.

Where the windows of the basement, as shown at *m*, are brought down close to the pavement, it is absolutely necessary that the pavement be sloped away from these windows with considerable pitch, not less than 1 inch in 1 foot, as otherwise the water is likely to lay against the window sill or run under it, causing it to rapidly decay, the capillary attraction of the window frame drawing up the water.

29. Window Openings.—Referring to Fig. 9, which shows the face view of a bay of the wall illustrated in section in Fig. 7, the details of the several window openings in the walls may be studied. The basement windows are independent frames with double-hung sash, a rowlock brick arch supporting the brickwork over the window head. In the practice and design of window heads for mill buildings, it is usual to make the radius of the window head equal to the width of the reveal. In this instance, the distance across the opening is 4 feet 3 inches, and the radius of the arched head is the same dimension.

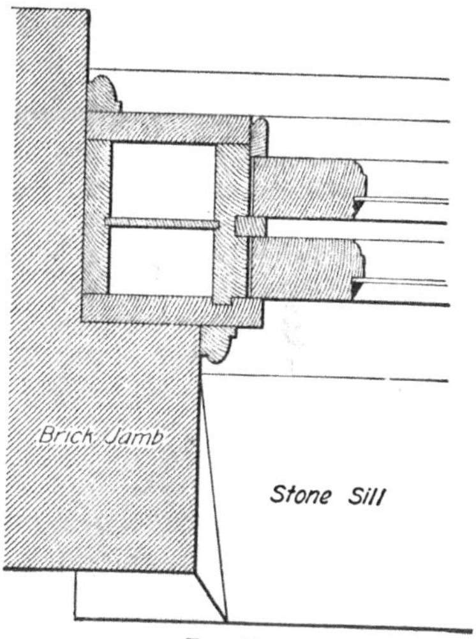

FIG. 10

The windows throughout the balance of the building are twin windows, double hung, and the construction of the window frame and sash is shown in the drawing. This frame is what is known as a **reveal frame,** and is built in as the brickwork progresses.

Sometimes the frame is slipped in from the back, as shown in Fig. 10, and when this is the case the work can be carried along without waiting for the window frames.

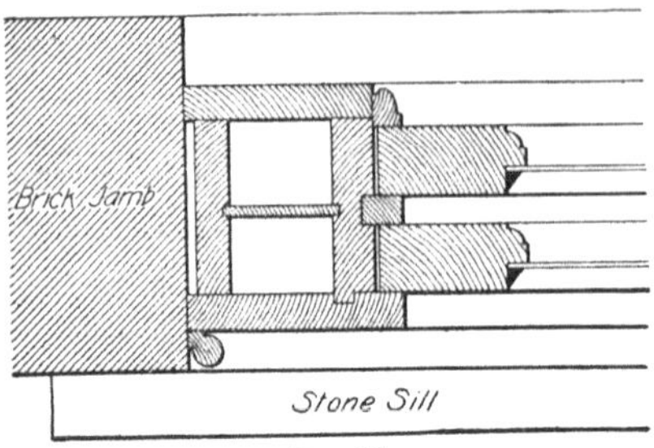

FIG. 11

As distinguishable from the reveal frame, there is the **plank-frame construction**, which is not built into the brickwork, but is built up as shown in Fig. 11. When it is desirable to have the central mullion *a*, Fig. 9, as narrow as possible, the box construction indicated on the drawing is done away with and the window is hung by means of overhead pulleys, the weights operating in the boxes at the sides.

STANDARD SLOW-BURNING CONSTRUCTION

30. A type of factory construction more usual than that previously described is illustrated in Fig. 12. In this illustration, it will be noticed that the main girders bear on wall pilasters, and the spandrel filling between the pilasters is kept as thin as possible. The usual reveal window frame is used, as shown at *a*, and the soffit of the arch over the window openings is checked at the head of the opening to provide a wind and water stop as at *b*. In this construction, which is probably the best, though it does not possess the advantage of giving the maximum amount of window space, and, consequently, light in the building, a rowlock or bonded brick arch is used over the window frames. By means of this construction, either the window frame may be built in place, or the windows may be slipped in from the back against a rabbet formed in the brickwork. The arch over the window head is indicated at *c*.

31. Floor Construction.—The floor construction consists of heavy timber girders, no dimension of which may be less than 6 inches, as otherwise it would not comply with the requirements of slow-burning construction. The floor planking consists of 3- or 4-inch tongued-and-grooved spruce, or yellow-pine planking, planned on the under side, and

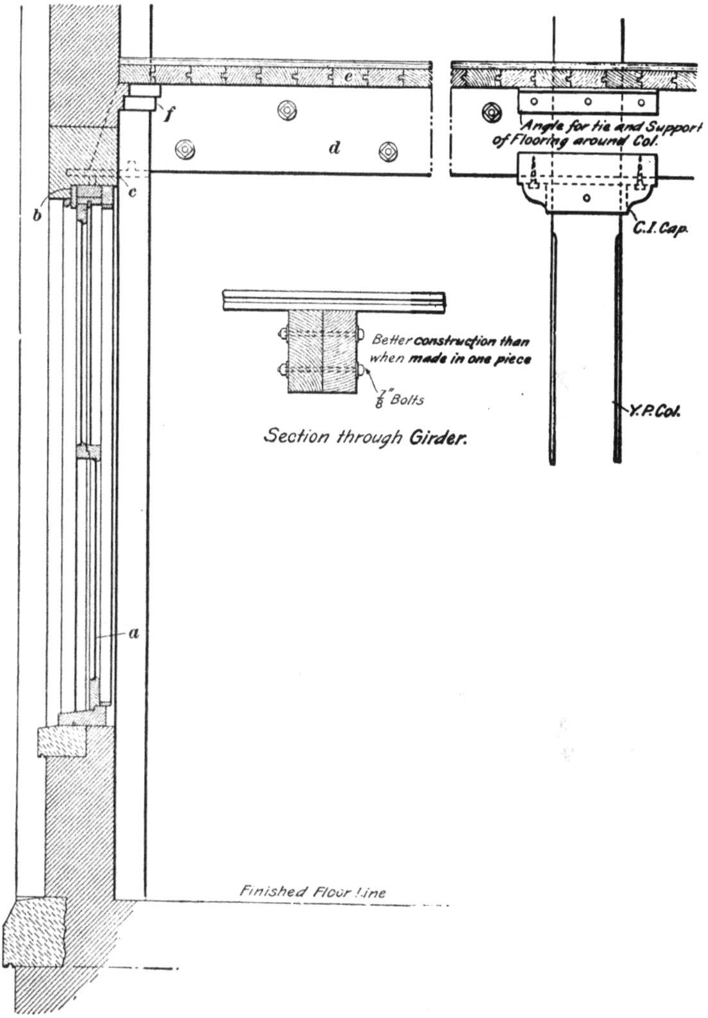

Fig. 12

thoroughly spiked to the girder. Planking of the former thickness may be used for clear spans as great as 8 feet, while the latter thickness may be used for up to 10-foot or even 12-foot spans, if the loads are light. The girders are indicated at *d*, and the floor planking at *e*. Usually the girders, in order to obtain the requisite strength, are made of long-leaf yellow pine. On the top of the spruce planking is placed a finished maple floor. This floor is made from either 1-inch maple, which finishes as $\frac{7}{8}$ inch, or $1\frac{1}{4}$-inch maple, which finishes as $1\frac{1}{8}$ inches, in thickness. Neponsett sheathing paper, or deadening felt, is placed between the spruce planking and the finished maple flooring for the purpose of preventing dust from percolating through. This sheating paper or felt is sometimes made waterproof to prevent leakage due to water used for fire-extinguishing purposes.

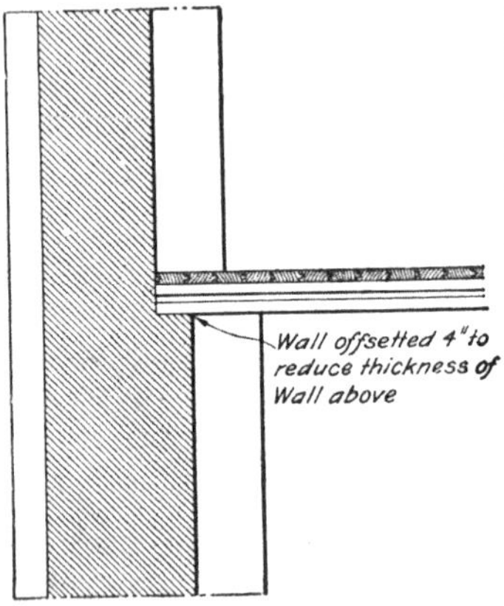

FIG. 13

Frequently, the brickwork is corbeled out, as indicated at *f*, in order to form a fire-stop between floors, or at least to prevent an open joint at this place. Where the walls are offsetted, as shown in Fig. 13, there is no need of corbeling out, for the offset in the brickwork can be made to form the fire or dust stop.

32. Where heavy yellow-pine girders bear on brick walls, it is usual to obtain the requisite bearing area by the use of cast-iron bearing plates, as indicated in Fig. 14 (*a*), (*b*), and (*c*). In (*a*) is shown an ordinary flat plate that has an area figured so that the load on the brickwork will not exceed its safe stress, which for brickwork laid in lime-and-cement mortar is about 150 pounds per square inch, while for brickwork laid in cement mortar, it is in the neighborhood of 200 pounds per square inch. This plate is usually cast with a lug on the back, as at *a*, to be built in the brickwork, and dowel-pins, or a lip, as at *b*, over which the girder is fitted, or notched. By this means, a tie to the wall is obtained. There is difficulty, however, in using such a connection, for the carpenters on the job frequently miscut their beams, so that the notchings or borings at *b* do not come where they should, and to remedy the defect, the notchings, or borings, are cut or gouged out, so that frequently the pin or lip at *b* is not brought to bear against the timber.

A more practicable bearing plate is illustrated in Fig. 14 (*b*). Here, instead of providing dowels, or a lip, to set into the girder, the top of the plate is cast with teeth, as indicated at *c*. While these teeth tend to destroy fibers at the bottom of the beam, they nevertheless sink into the timber, creating great friction, and thus accomplish a tie to the wall fully as efficient as a dowel-pin, or lip, let into the timber would be.

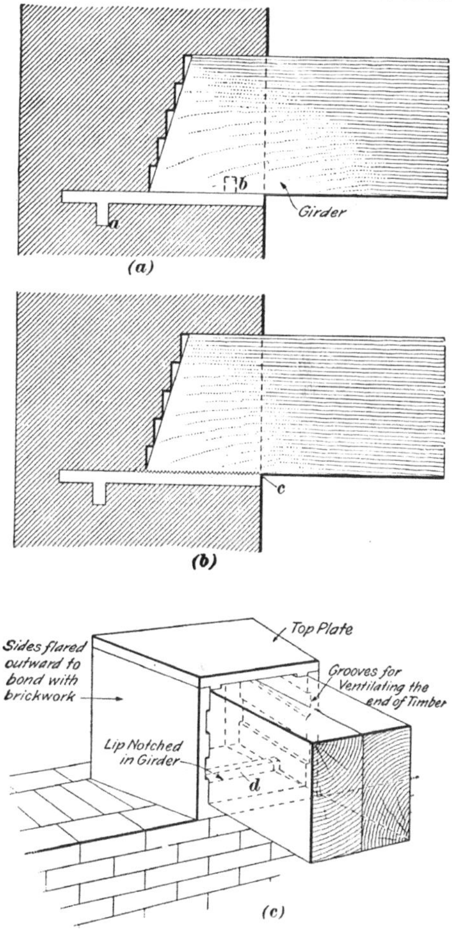

Fig. 14

Probably the most common form of bearing plate is that illustrated in Fig. 14 (*c*), which is known as the Goetz-Mitchell bearing box. This is usually built flared, as indicated in the illustration, so that when built into the brickwork it will have a hold in it, and the timber acts as a tie by being notched over the lip, as at *d* in this figure. These Goetz-Mitchell boxes are generally provided with a plate that sets on top of them, on which the brickwork may be built, and not infrequently the sides of the boxes are grooved so that the ends of the girders are ventilated.

33. Window Heads.—In Fig. 12 was shown a form of window head that is the best for strength, but possesses the disadvantage of lowering the top of the window, thus cutting off light to the room, which is a serious objection where the room is wide, or where it depends on the windows in one side for lighting the entire floor area. In order to keep the window head up near the under side of the floor construction, an I beam, lintel, or some similar form of support for the brickwork over the head that takes up little room, must be employed. A construction using shallow I beams is illustrated in Fig. 15. Here the window head is directly beneath the rough flooring; and while the outside face of the window is formed with an arch, the brickwork above the window head is supported on shallow I beams. This figure illustrates a section through the wall extending parallel with the main girders, a bearing being obtained for the floor planking by bolting to the I beams a bearing strap *a*.

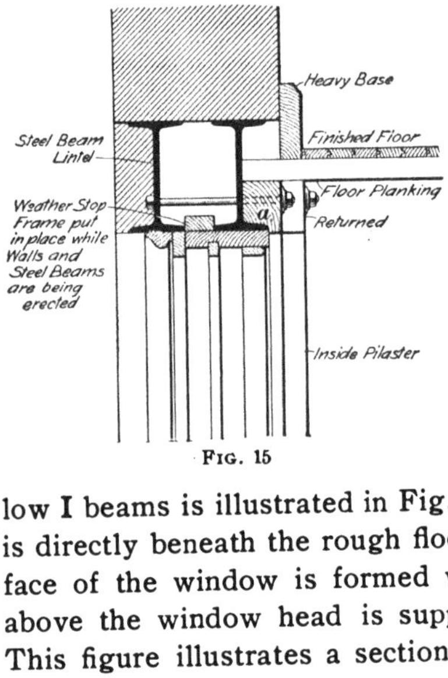

FIG. 15

This construction would not be permitted in some of the larger cities, as the building laws require that all steel beams supporting brickwork must be fire-proofed. Consequently, a steel lintel of this construction would have to be surrounded with concrete, and the window head dropped somewhat to allow a bearing for the floor planking, or some other form of construction adopted.

FACTORY BUILDINGS OF REINFORCED CONCRETE

34. Within the last few years, the cost of the best Portland cement has been so materially reduced that concrete has become an available material for the construction of facto-

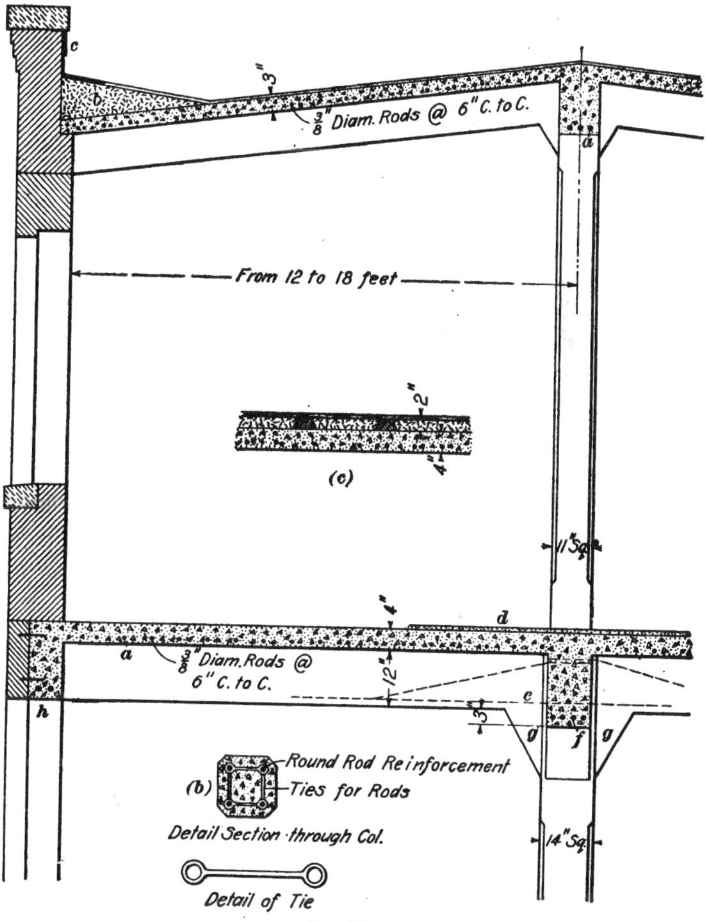

FIG. 16

ries. Unless used in great masses, however, it has not the strength to support the necessary floor loads without the use of steel reinforcement. As explained in *Design of Beams*, the fibers on the bottom of all beams subjected to transverse stress are in tension, and while concrete has considerable resistance to compression, it offers comparatively little to tensile stress. It is therefore necessary to reinforce the lower portion of all beams and floor slabs as indicated at *a*, Fig. 16.

35. Advantages of Reinforced Concrete.—In Fig. 16, the details of a typical reinforced-concrete factory building are illustrated, and a building of this character may be constructed for a cost of from 10 to 15 per cent. greater than the ordinary slow-burning type of building. Besides, this construction possesses the advantage of being practicable for long spans and heavy loads, whereas in buildings of the slow-burning type, owing to the fact that the size of the wooden beams is limited to the available commercial timber, it is frequently impossible to design floors with girders of large spans for floor loads of over 250 pounds per square foot. While this is a heavy load, it is too light for some classes of work, such as occur in printing houses and lithographing establishments where heavy stones are used and stored. The floor loads in such buildings sometimes amount to as much as 300 or 400 pounds per square foot, while it is not unusual to find the load on floors in warehouses amounting to as much as 500 pounds per square foot.

36. Strength of Concrete Columns With Steel Cores.—In the building shown in Fig. 17, it will be noticed that the columns are reduced in size in the lower floors, increased in the middle portion of the building, and reduced toward the roof. The reduction in the columns *a* and *b* is due to the fact that these columns are reinforced with a steel core composed of structural shapes riveted together, angles usually being employed for this purpose. In proportioning such columns, it is good practice to figure on the ultimate safe unit compressive stress of the steel without considering the reduction made by the usual column formula, but to neglect, in the consideration of the strength of the column, the resistance of the concrete surrounding the steel core,

To illustrate, if the sectional area of the steel reinforcements in these columns equals 20 square inches, and a safe unit fiber stress of 16,000 pounds is assumed, the safe strength of the column will be 320,000 pounds.

Above the second floor, the columns are made much larger, for here there is less steel reinforcement, and it is necessary to figure on the safe bearing strength of the concrete.

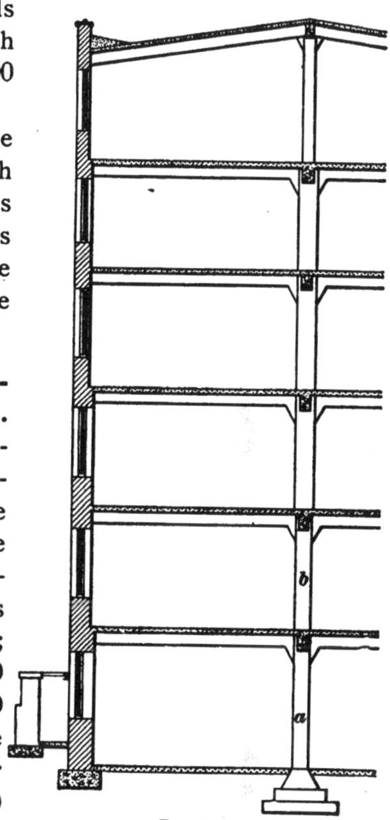

FIG. 17

37. Strength of Reinforced-Concrete Columns. In proportioning reinforced-concrete columns, it is customary among conservative engineers to figure the safe strength of the concrete-column section at 500 pounds per square inch of section; that is, if the column is 20 inches square, its area is 400 square inches, and its safe strength at 500 pounds per square inch will be 200,000 pounds. In the top floor, it is seldom advisable to use concrete columns less than 10 inches square, though at this dimension they generally possess several times the requisite amount of resistance.

All columns in reinforced construction generally have embedded in them $\frac{3}{4}$-inch to 1-inch round steel rods, tied together with round iron binders, or bar-iron straps as indicated in Fig. 16 (*b*).

38. Floor and Roof Construction.—In considering the floor and roof construction of buildings built of reinforced concrete, it will be noted from Fig. 16 that the roof slab is made 3 inches in thickness. Such a slab made of good concrete, reinforced with $\frac{3}{8}$-inch steel rods, spaced 6 inches from center to center, will carry the usual roof loads for spans up to 7 feet in the clear.

In forming the gutter for such roofs, as indicated at *b*, the gusset is made by filling in with cinder concrete. Usually cast-iron eave boxes are embedded in the concrete, and these in turn connected with inside rain conductors.

The beams supporting the roof, when the span is from 12 to 14 feet, are made about 12 inches deep and 8 inches wide, while the girders, also constructed of reinforced concrete, are usually made about 3 inches deeper and 11 inches in width.

In order to make the roof impervious to moisture, a covering of felt and slag is commonly employed. This slag joins the parapet wall with the usual tin flashing and counter flashing, as at *c*, though copper is recommended for best work.

In the floor construction of reinforced-concrete factory buildings, the slabs forming the floor panels are made not less than 4 inches in thickness, and seldom over 5 inches, with a 1-inch finish coat of cement besides, if this character of finish is desired. Such a floor slab is shown in the construction at *d*, Fig. 16, while the wooden floor construction is shown in Fig. 16 (*c*). Here the structural feature of the floor is a 4-inch concrete slab upon the top of which is placed 2″ × 3″ beveled hemlock sleepers, the space between these sleepers being filled with cinder concrete, and the floor finish obtained by laying 1-inch tongued-and-grooved maple floorings.

39. Reinforced-Concrete Beams and Girders.—The depth of the beams and girders in reinforced-concrete construction varies, of course, with the span and loads to be supported. Their width enters little into the strength, and they may be made as narrow as possible in order to cover the reinforcing steel. It is the best practice to make beams and girders of the same width, for then the process of forming the molds is greatly simplified and the cost reduced.

In placing the reinforcement in the concrete, it should always be at least 2 inches from the outside surface, for a distance less than this is considered inadequate fireproofing. In order that the reinforcing metal *e*, Fig. 16, may enter over the top of the reinforcing metal at *f*, it is usual to make the secondary girders, or beams, 3 inches less in depth than the main girders. To stiffen the building, brackets are customarily introduced between the column and girders, as illustrated at *g*. These brackets tend to greatly increase the rigidity of the connection and shorten the span of the girder somewhat.

40. Construction at Window Heads.—Where it is necessary to have the window head near the top of the

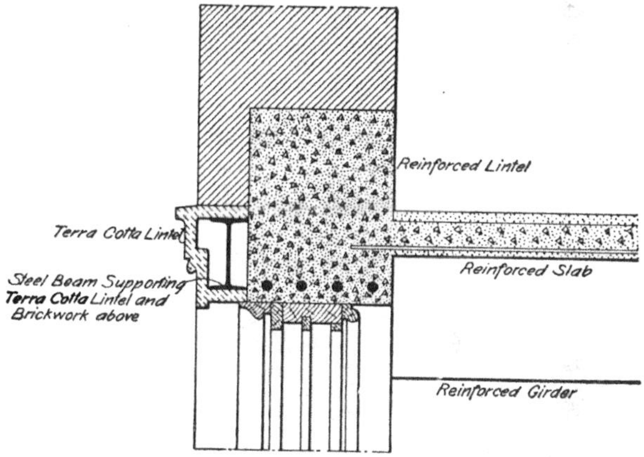

FIG. 18

ceiling, reinforced-concrete construction lends itself readily to the requirements of this condition, for even where girders are supported over the window head, the construction may be followed out, as indicated at *h*, Fig. 16. Where it is desired to have the window head raised still higher, a construction similar to that shown in Fig. 18 may be used. In this case, however, care must be taken to have the girders bear on the piers between the windows, and to have no intermediate beams.

41. Column Footings.—With factory buildings of more than five or six stories in height, great pressure is transmitted to the soil from the base of the bottom column, and as it is

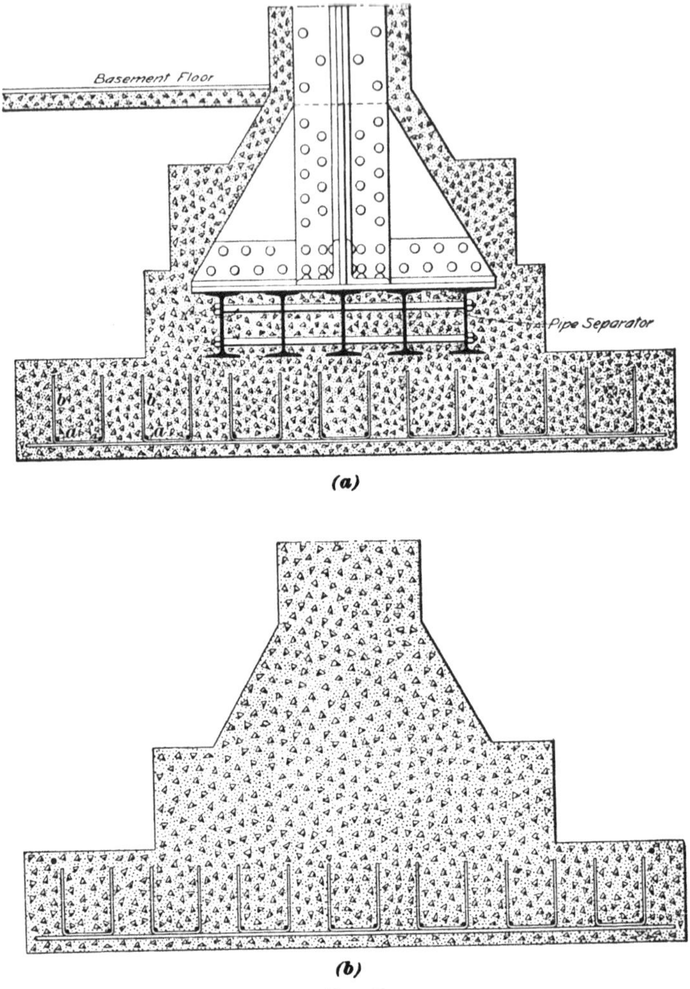

(a)

(b)

FIG. 19

necessary with soils of even fairly good bearing capacity to have footings beneath the piers supporting columns of from 6 to 10 feet square, adequate means of providing these footings must be obtained. In Fig. 19 (*a*) and (*b*) are shown two types of footings for concrete columns. In (*a*) is indicated a reinforced-concrete column with a steel core. In such an instance, all the load is transmitted by the steel core through its angle plates and webbing at the foot to grillage beams. These grillage beams are, however, not made sufficiently large to transmit the load to the soil, but merely to distribute the load on the bed of concrete. The spread portion of the footing is reinforced with steel rods *a*, *a* crossed each way, and longitudinal shear is taken up in the footing by means of stirrups *b*, *b*. This is the usual type of footing construction under reinforced-concrete factory columns.

Where, however, the column is not reinforced with a steel core, but is merely a pier, footings may be designed as illustrated in Fig. 19 (*b*). Here the base of the column is enlarged in order to better distribute the load on the several steps of the footing, and where the bottom step has a considerable overhang, it is reinforced with steel rods and stirrups, as indicated.

42. Detail of Slab and Girder Reinforcement.—In the previous article, the general construction of the floors and column supports of a factory building was explained. By referring to Fig. 20, it will be shown how the girders and beams are reinforced with the steel bars. In this figure, a plan is indicated at (*a*) and an elevation at (*b*). The rod reinforcement of the slab is shown in the plan at *a*, *a*. It will be noticed that over every other beam these rod reinforcements lap, or break joints, and that some additional tie or reinforcement is placed over the girders, as indicated by *b*, *b*. These latter rods tend to tie in the floor slabs still more rigidly than can be accomplished with their individual reinforcement.

Referring to the elevation (*b*), it will be noticed that all the reinforcement of the beams is not usually carried along the lower portion of the girder for its entire distance, but that some of the reinforcement is bent up at a point about one-quarter of the span from the abutment, in the form of a camber rod. By arranging the reinforcing rods in this manner, an additional stirrup action, or tie, to the girder supports is provided, and the oblique section made by a horizontal line passing through these rods tends to provide additional resistance to the horizontal shear in the beams and also provide for negative bending moment produced in

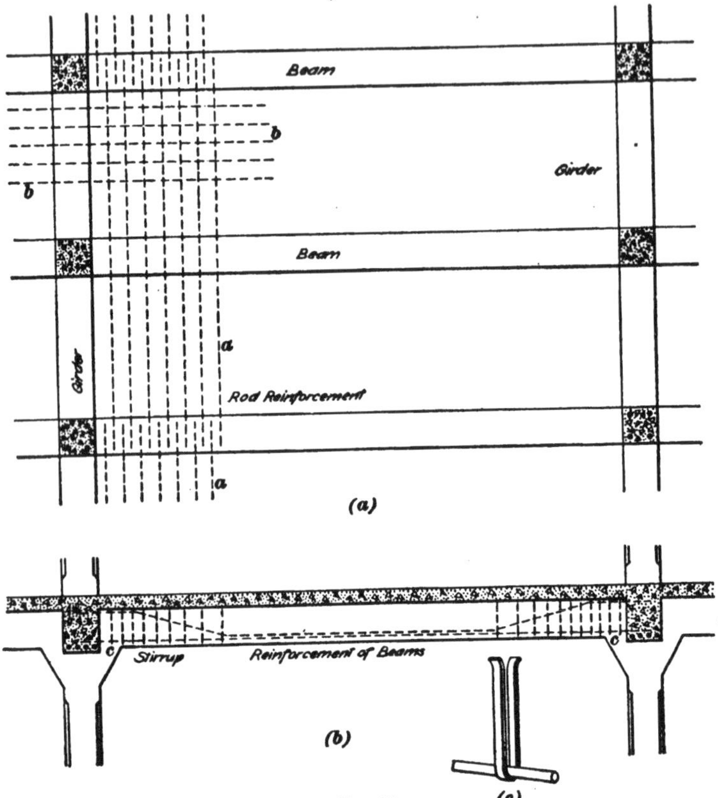

FIG. 20

the beams near the support. To further provide for this, shear stirrups are placed closer together, toward the abutments, as indicated at *c*, *c*. These stirrups are ordinarily light pieces of bar iron bent in a **U**-shape, and sometimes bent around the rod reinforcement, a detail of this stirrup being shown in Fig. 20 (*c*).

STEEL-FRAME MILL BUILDINGS

43. There is a type of building which, while not distinctly mill construction as usually understood, is frequently used for one-story buildings, such as rolling mills, cement works, machine shops, foundries, rail yards, and buildings of this class.

The essential feature of these buildings is a steel-roof truss supported on steel columns, the columns being braced both to the truss and longitudinally of the building. It is usually the purpose in the design of such buildings to neglect everything but the necessary stability and the first cost. The steelwork, consequently, is of the lightest possible construction, usually designed for a unit fiber stress of from 18,000 to 20,000 pounds, and the covering of the sides of the building, together with window details, etc., is made only sufficiently good to keep out the weather.

44. Material for Roof Covering.—The roof covering of this class of building is either of slag on 2-inch spruce plank, spiked to nailing strips bolted on to steel purlins from beneath, with lagscrews, or of slate laid on 1-inch or 2-inch sheathing boards. Even galvanized iron is used for the roofing of some of the cheapest class of buildings, especially those which, owing to the process of manufacture, are subjected to a high temperature.

45. Construction of Sides of Building.—The sides of these buildings may be covered with either expanded-metal lath on metallic furring strips, plastered inside and out with cement mortar so as to form a fireproof and rigid screen wall about 2 inches in thickness; or, the walls may be 9-inch or 13-inch brick walls built part way up the height of the columns and leaving the columns exposed on the face; or, corrugated galvanized iron lapped 6 inches and secured either by riveting to metallic supports or nailed to wooden studding secured to the steel frames. Of these constructions, probably the first is the most expensive and also the most satisfactory.

46. Partially Supported Steel-Frame Building. In Fig. 21, there is designated a type of construction that

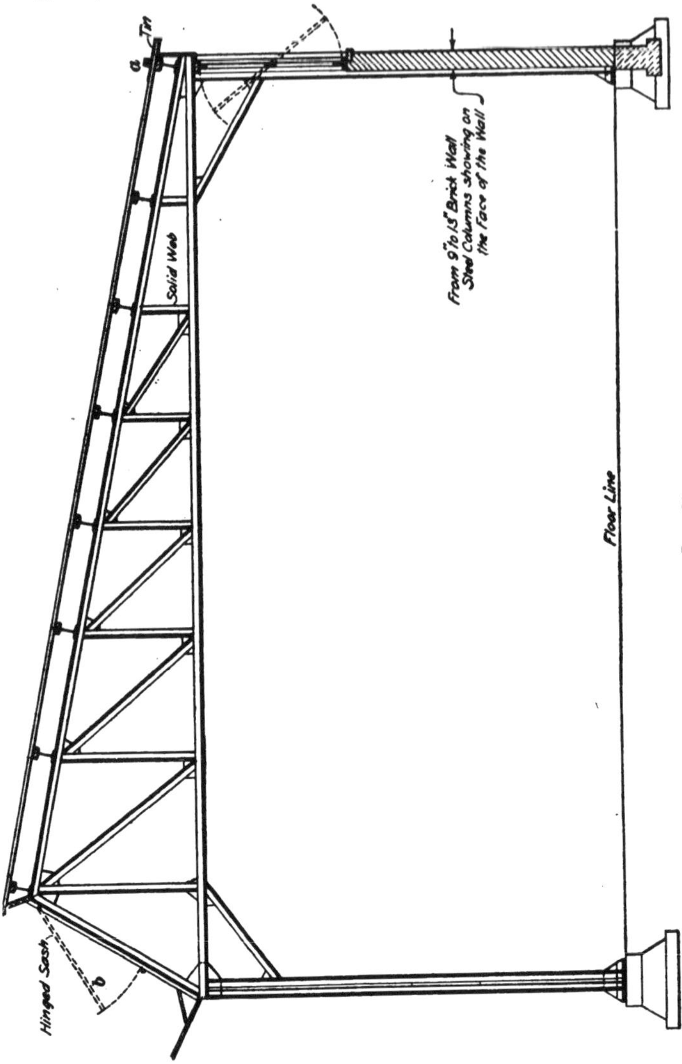

FIG. 21

may be built for about $1 per square foot of the area covered. This consists of steel **I** beams, or angle-and-plate columns, used for column supports carrying the usual angle iron steel-roof truss. The roof is sheathed with 2-inch spruce tongued-and-grooved planking, covered with a good quality of roofing felt and slag, with a stop-gutter *a* at the edge. Owing to the fact that the steel columns are supported in a direction of their minimum radius of gyration by means of the brick walls, they can be made very light. The building illustrated has what is known as a *saw-tooth roof*. By this means, light is obtained on the side next to an adjacent and higher building by means of a sash *b*. This sash is usually made hinged or pivoted, to provide the necessary ventilation.

47. In Fig. 22, there is illustrated, diagrammatically, the framework of a one-story skeleton-construction building.

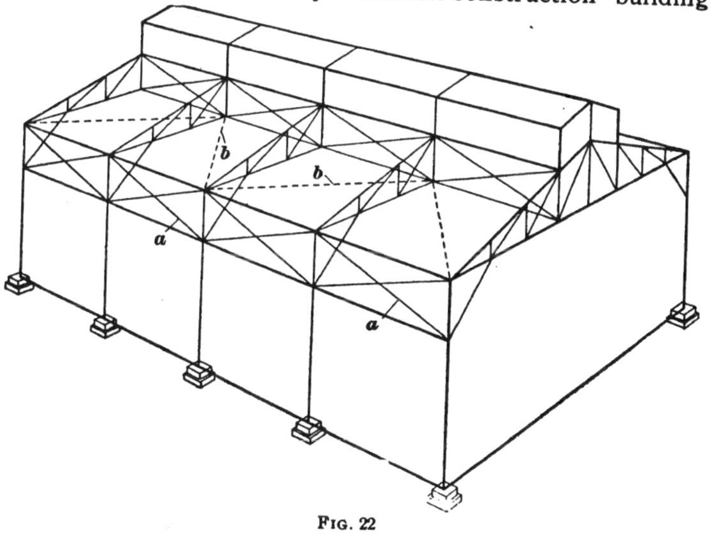

FIG. 22

In the design of all such buildings, where there are no end gable walls, the several columns and trusses must be braced diagonally, as indicated at *a*, *a*, and frequently it is necessary to introduce a secondary system of horizontal bracing from one panel point on the lower chord to another, as indicated at *b*, *b*.

In placing galvanized ironwork on the sides of steel-mill buildings, it is best to construct the necessary framework between the main supporting members of the building of light angles, or tees. These should be furnished punched with $\frac{3}{8}$-inch or $\frac{5}{16}$-inch holes, to which the galvanized iron may be riveted, it being best to mark the galvanized iron in the field and punch it there. This may be done without much difficulty with the usual light gauge used for this purpose. It is sometimes necessary with this construction to flash around the window and door heads with IX tin.

DETAILS OF MILL CONSTRUCTION AND DESIGN

STRUCTURAL FEATURES

BEAM CONNECTION TO GIRDERS

48. In factory construction, the headroom is seldom available to support beams on the girders, as indicated in Fig. 23 (*a*). It is usually necessary, in order to cheapen the construction of mill buildings, to keep the distance between the clear headroom and the finished floor level to the very minimum, and consequently the tops of the beams are most always brought flush, or nearly so, with the top of the girder.

A common construction is to use some of the various forms of wrought-iron hangers, as shown in Fig. 23 (*b*). The type of hanger shown is a single stirrup, and is probably the best of any on the market; where beams enter the girder on both sides, the hanger is designed double. While it is popularly supposed that this hanger would readily fail by the bending of the metal at *a*, it is usually proportioned to safely carry any reaction imposed under ordinary floor loads. This hanger is obtained stamped out of steel plate or formed from bar iron.

49. Where it is not desirable to use wrought-iron or steel hangers, a simple and inexpensive form of construction may be adopted as that shown in Fig. 23 (*c*). Here the beam *a* is supported on a wooden strip *b*, which extends the full length of the girder, and is bolted near the bottom with through bolts. Such a construction provides sufficient strength for the support of the average factory floor, but its strength is difficult to figure with any degree of certainty, and some surer form of connection is generally considered preferable. In all instances, it is good practice to tie together the opposite floor-beams butting on a girder by means of an iron dog, or tie-plate, *c*.

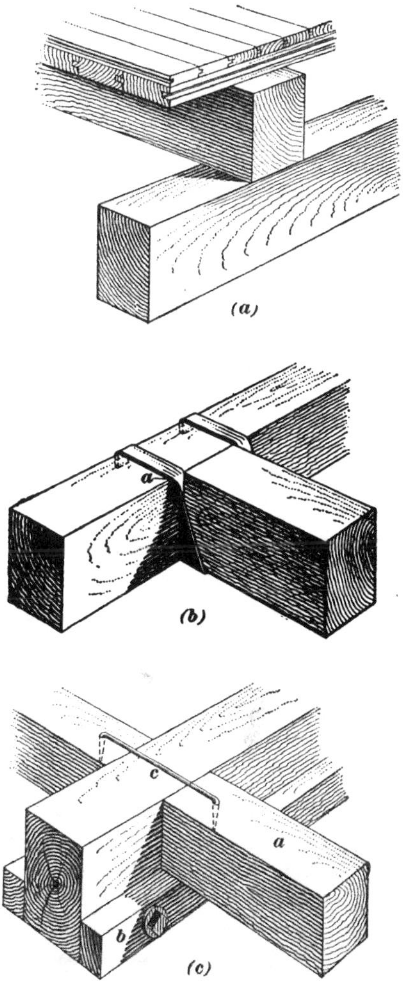

FIG. 23

50. In Fig. 24 (*a*), (*b*), (*c*), and (*d*) are indicated other methods of supporting the secondary floor-beams on main girders in the construction of factories. In Fig. 24 (*a*) is shown an I-beam girder supporting heavy timbers of a floor of slow-burning construction. It is always necessary in this construction to bring the top edge of the timbers above the upper flange of the I beam, and to span the space *a* thus created with a piece of timber for a tie and for the support of the floor planking. By providing this space between the ironwork and the wooden tie, any shrinkage that may occur in the secondary timbers will not cause the floor to ride on the top of the steel beam and thus make a ridge evident in the finished floor at this place. The timbers forming the secondary girders may either be supported on angle-iron brackets, or on angle irons extending the entire length of the girder. The latter method is only pursued when it is necessary to keep the end of the timber a few inches away from the steel beam, and the angle, consequently, being subjected to a greater bending moment, must have more resistance by increasing the width of the section of the bracket.

Sometimes, the secondary beams are supported on double

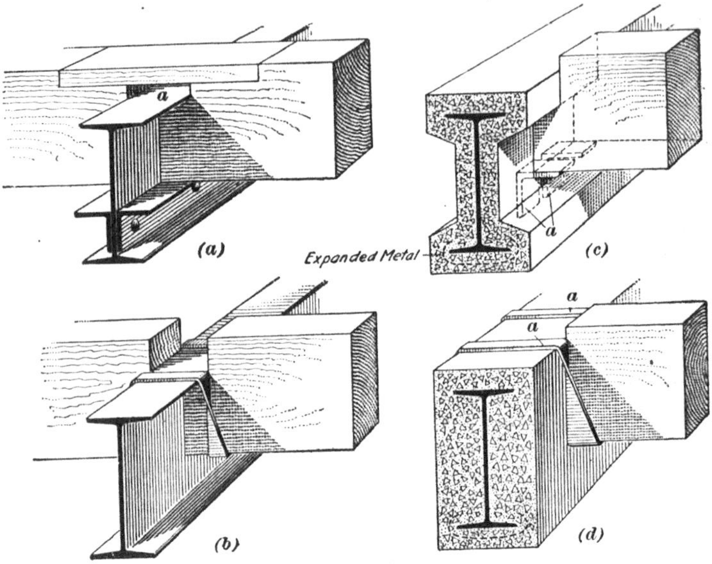

FIG. 24

stirrup hangers, as shown in Fig. 24 (*b*). When it is not desired to use steel beams, resort is frequently had to flitch-plate girders. They are, however, held in some disfavor by the building departments of the several cities, who do not consider that the combined strength of the timber and metal can be taken, and will only permit the strength of either the timber or metal to be used.

51. The building departments of several of the large cities stipulate that buildings of the second class, which includes factory construction, shall not have steel girders that are not fireproofed supporting brick walls or floors. When this construction is required, the secondaries must be supported as in Fig. 24 (*c*). In this view is two angle brackets riveted or bolted to the steel beam, and extending through the concrete for the support of the wooden beams. While there is some danger of heat being transmitted to the beams through the projecting ends of these brackets, nevertheless it is considered better construction than that shown in Fig. 24 (*d*), where stirrups are used over the concrete fireproofing. In this latter construction, there is a liability of the stirrup bending at *a*, *a*, and crushing the concrete beneath. Where the reaction from the end of the girder is great, this undoubtedly is likely to occur, and such stirrups should be provided with a bearing plate on top of the concrete, so that their bearing at the edge will be distributed over a considerable area.

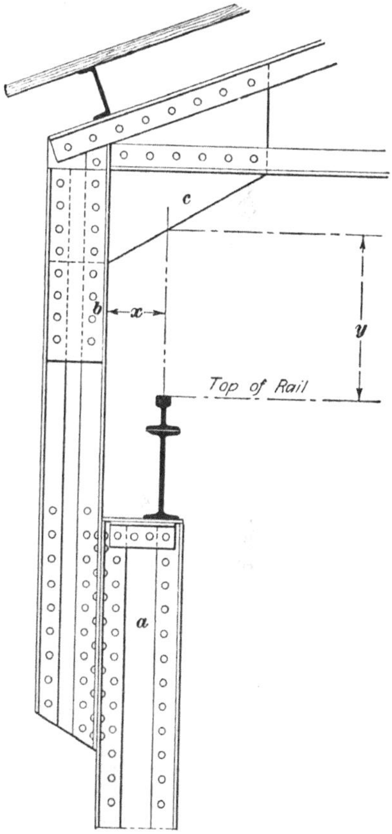

FIG. 25

TRAVELING-CRANE LOADS

52. Planning for Traveling Cranes.—In designing factories or mill buildings in which traveling cranes are to be installed, it is important to observe that the track of the crane can be properly supported, and also that there is sufficient headroom under the floor or roof construction to permit the trolley of the crane and the traveling mechanism of the crane girder to move underneath.

In Fig. 25, there is shown the upper portion of a steel-mill building. The columns *a* support the girder carrying the runway of the crane. A convenient means of supporting the roof is to splice to this column a similar column *b*, which

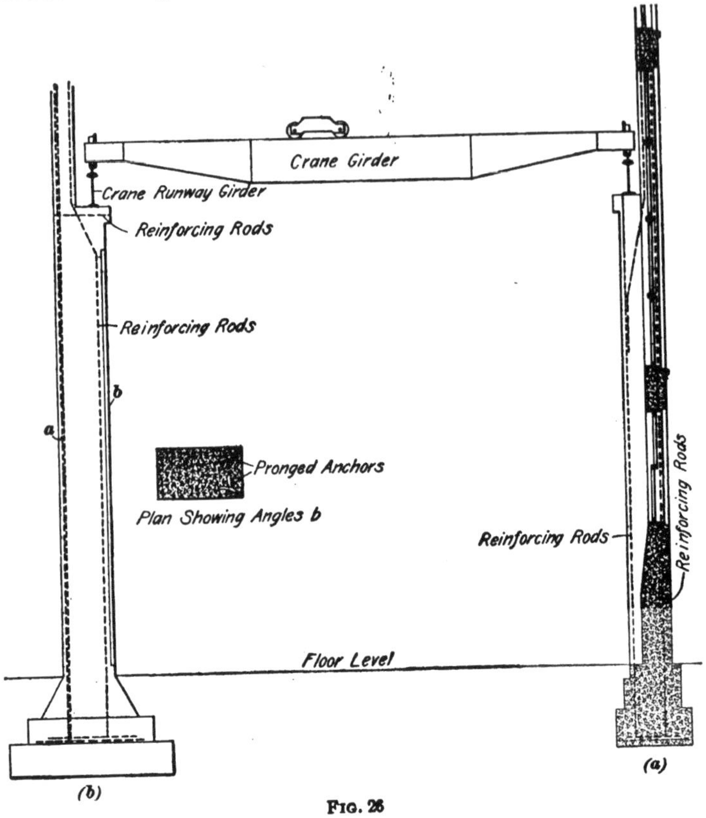

FIG. 26

is incorporated in the design of the roof truss and rigidly braced with the truss by means of a knee brace at *c*. In the design of such a building, it is very important to determine the distances *x* and *y* required by the makers of the traveling crane. These distances *x*, *y* depend on the size of the crane, that is, whether it is designed to carry 5, 10, 15, or more tons. Usually from 9 to 12 inches is sufficient for the measurement *x*, while the measurement *y* varies from 5 to 8 feet.

53. Cranes Supported on Reinforced-Concrete Walls.—Frequently, in the latest types of construction, the runway for the crane is supported on reinforced-concrete walls, which construction is shown in Fig. 26 (*a*). It will be observed that the pilasters supporting the crane are strongly reinforced in all directions from which stresses are likely to be created from the eccentric load imposed by the crane track.

Where cranes are supported on reinforced-concrete columns, as in Fig. 26 (*b*), it would be good practice to put additional rods in the far side of the column as at *a*, in order to supply a greater resistance to bending, and thus counteract the effect of the eccentric load produced by the reaction from the crane track. Where cranes handle heavy rails or cumbersome material that might, by swinging, impose a blow on the reinforced-concrete columns, it is good construction to protect the edge of the columns with an angle iron as indicated at *b*. This angle iron may be fastened in the forms and anchored by means of pronged anchors back into the concrete when it is tamped.

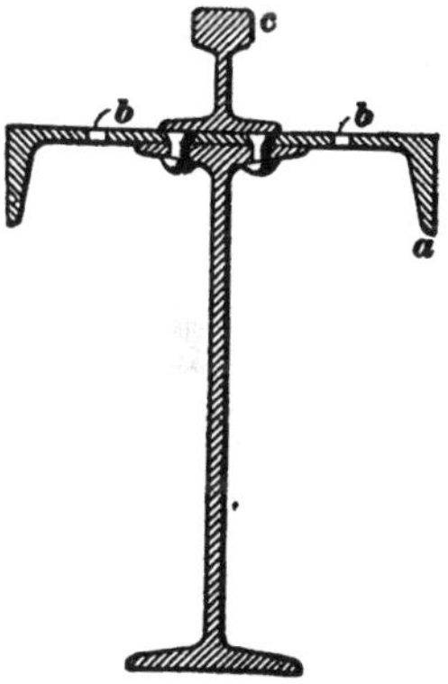

FIG. 27

54. Detail of Track Construction.—Many crane failures have been due to the spreading of the track between supports. It is better, therefore, to supply considerable lateral rigidity to the beam supporting the track or traveling crane. Where loads are heavy and plate girders are used for the runway tracks, the flanges of the girder are sufficient for this purpose. Where I beams are used, however, for the support of the crane track, it is good practice to place on the top of them and rivet with countersunk rivets, spaced about 18 inches apart on each flange, channel irons as indicated at *a*, Fig. 27. By means of these channel irons, which are drilled with open holes *b*, *b*, the rail *c* may be readily clamped in place by means of wrought-iron clips and bolts, and the rails nicely alined and adjusted by wedging between these clips and the track.

55. Maximum Stress on Track Girders.—The principal calculation for the construction of the runway of cranes exists in determining the maximum bending moment. The maximum bending moment on a runway girder occurs when the wheels of the traveling crane are in the position indicated in Fig. 28. It will be noticed that the center of the

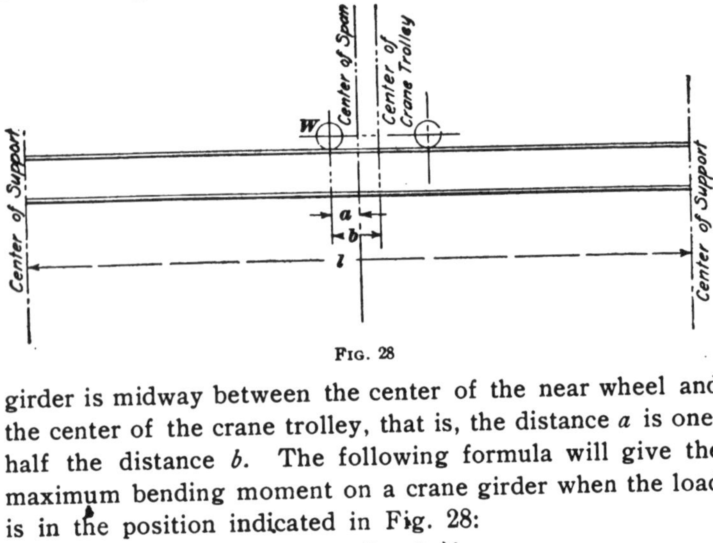

FIG. 28

girder is midway between the center of the near wheel and the center of the crane trolley, that is, the distance *a* is one-half the distance *b*. The following formula will give the maximum bending moment on a crane girder when the load is in the position indicated in Fig. 28:

$$M = \frac{w(l - 2a)^2}{2l}$$

in which M = bending moment, in inch-pounds;
w = load on one wheel of crane, in pounds;
l = span of girder from center to center of support, in inches;
a = distance, in inches, marked in Fig. 28.

In order to illustrate the application of this formula, assume that the wheel load w equals 10,000 pounds; that the distance from center to center of supports of the runway girder is 15 feet, or 180 inches; and that the distance *a* is 12 inches. By substitution,

$$M = \frac{10,000 \times (180 - 2 \times 12)^2}{2 \times 180} = 676,000 \text{ inch-pounds}$$

From this bending moment may be found, by the methods given in *Design of Beams*, the proper size girder to use.

THE POWER PLANT

BOILER ROOM

56. Locating the Boiler Room.—The ideal location for the boilers of a factory or an industrial plant is in a separate building, which may be denominated as the **power house,** and which may include as well, the installation of the engines, dynamos, and other machinery necessary for the generation of power and its transmission. More frequently, however, the ground is not available for the erection of a separate building for the power plant, and it becomes necessary to install the boilers and engines in the factory itself. The location usually selected for these vital features of the mill is the basement, and the arrangement of the boilers and engines must be carefully considered in the designing of this portion of the building.

57. In laying off the space to be occupied by the boilers, the probable growth of the manufactory must be provided for by arranging ample space for the installation of additional boilers.

It is best in arranging the boilers, to face them toward the available coal supply, which is usually a coal bunker, vault, or bin, but in no instance must the front of the boiler be nearer to a wall than the length of the boiler tubes, unless special arrangements are made, for this distance must be allowed in order to draw any defective or damaged tubes and replace them with new ones. Also, by arranging the boilers thus, the fireman has a minimum amount of carriage for the coal.

58. Coal Storage.—In designing the coal vaults, or coal storage, their contents should be figured to allow for 1 or 2 weeks' coal supply, and as much more as is possible, to carry the plant over periods of existing coal shortage due to strikes or interrupted traffic from bad weather or other cause. In calculating the amount of space required for coal storage, it is sufficient to multiply the number of horsepower generated by the boilers by 4, which is the approximate number of pounds of coal per hour for the generation of 1 horsepower. This result, again multiplied by the number of hours for which the boilers are run at their capacity, will give the quantity of coal needed per day, in pounds. The weight per cubic foot of coal varies from 80 pounds for soft coal to 90 pounds for hard coal, so that by dividing the number of pounds by these quantities the cubic feet of coal required per day is obtained. The bins may then be proportioned for the number of days' supply which the judgment of the designer may assume as being necessary.

59. Ash Disposal.—Besides the consideration of the coal supply, some disposition must be made of the ashes from the boilers. Frequently, a bin is constructed of masonry, alongside of the coal supply, into which the ashes are dumped by means of barrows. In large plants, this bin can be emptied by means of an ash conveyer, or elevator, which will carry the ashes to the level of the street or railroad track and thence into a cart or car.

60. Planning the Boiler Room.—In locating the boilers in the boiler room, which should be done in the plans of the building, for it is not customary to cement the floor space covered by the boilers, and the cost of the building is thus reduced, a passageway not under 3 feet, and better 4 feet, should be left back of the boilers. This passageway is required in order to have access to the clean-out doors and the blow-off cocks. The ordinary horizontal return-tubular boiler, and some water-tube boilers, can be constructed in a battery, with as many boilers as may be desired in a row, especially when the passageway is left back of the boilers.

When setting other types of water-tube boilers, space should be left between each battery of two, for in these boilers, cast-iron doors are provided in the side walls for blowing the soot from the tubes, and access must be had through the side walls of the boiler for this purpose. It is therefore necessary in laying out the boiler space for boilers of this character to provide a passageway on one side of each boiler. In Fig. 29, a battery of return-tubular boilers is indicated, showing the clean-out doors for taking away the accumulation of

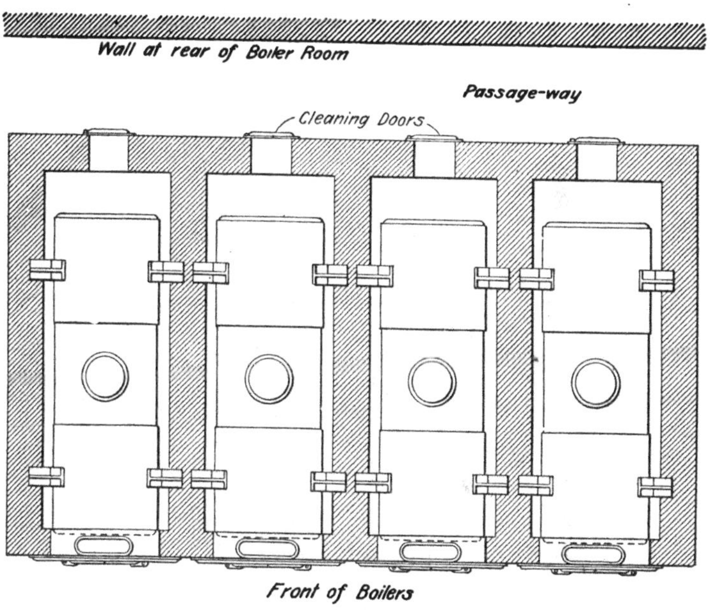

FIG. 29

soot and ashes that might be back of the bridge wall, through a passageway at the rear of the boilers. Some water-tube boilers are set in batteries of two, as the Babcock & Wilcox water-tube, land-type boiler, which is provided with the necessary clean-out doors, and doors for blowing the soot off the tubes in the side wall.

It is therefore necessary in laying out the boiler room of a manufacturing plant to consider carefully the character of the steam generator and study its requirements, so that it may be successfully operated and the proper spaces allotted.

61. Doorway to Engine and Boiler Room.—In the hasty design of buildings, it is frequently found that the size of the doorways is not sufficient to admit the boilers and machinery. This is a serious defect in the planning of a manufacturing plant, as it requires either the installation of the boilers and engines before the walls are entirely built,

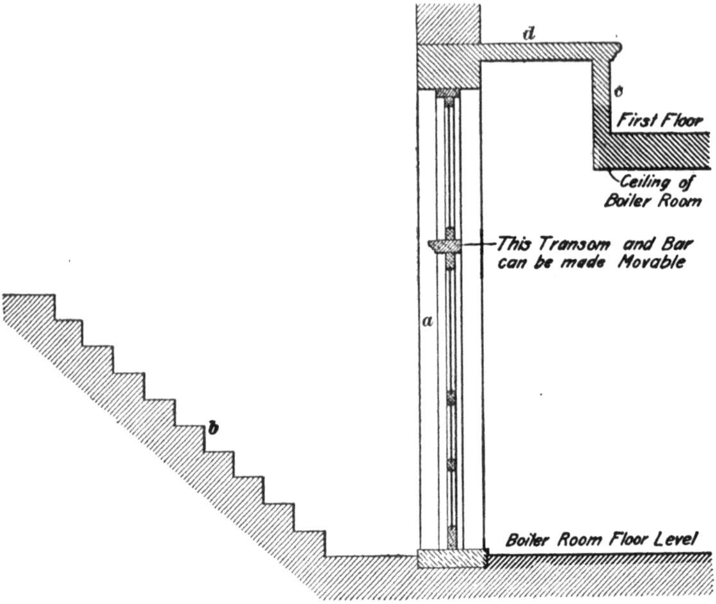

FIG. 30

or else the tearing out of brickwork and jambs in order to accommodate them afterwards. An expedient for the enlargement of the headroom of doorways and openings into the boiler and engine rooms that are in the basement, is shown in Fig. 30. Here, if the lintel of the doorway *a* is kept below the floor level, where it would ordinarily exist, the headroom of the doorway will be materially reduced, and considerable difficulty will be encountered in taking any large piece of machinery, or a boiler or steam drum, down the steps *b*. The doorway is consequently increased in height by the introduction of the bulkhead at *c*; while by this means the floor space above is slightly reduced, yet use can frequently be found for the ledge or platform frame at the top of the bulkhead, as at *d*.

62. Floors Above Boilers.—It is important in designing boiler rooms in factories to have the floor construction over the top of the boiler of incombustible material, and it is customary in the better class of buildings to provide a section of fireproof floor over the top of the boiler room. This floor construction may either be a brick arch supported on steel beams, or hollow-tile construction, though reinforced concrete is now finding favor in this purpose.

63. It is not altogether necessary that the boilers in a building shall be placed in the basement, though as this is usually the least valuable of the floor space it is the practice to so locate them. In some electric-light stations, and in large factories, boilers have been located on the first floor, and even in several instances on the fifth and sixth floors. The exigencies that demand the latter installation, however, must be great, for it can be readily seen that much power must be expended in lifting the coal, etc. to the boiler room.

CHIMNEYS

64. Dimensions and Capacity of Chimneys. Nearly all the factory buildings combine in their structure a power plant, not the least important feature of which is the **chimney.** There are two things to consider in the design of a power chimney—first, its capacity for providing the necessary draft and the conduction of the requisite volume of gases from the furnace or boiler, and second, its stability. The first requirement regulates its diameter and height, and these dimensions, together with its construction, determine also its stability.

Considering the first requirement, a circular flue is considered more efficient than a square one, because its inside

TABLE I

Diameter Inches	Height of Chimneys and Commercial Horsepower Capacity											Side of Square Inches	Effective Area Square Feet	Actual Area Square Feet
	50 Feet	60 Feet	70 Feet	80 Feet	90 Feet	100 Feet	110 Feet	125 Feet	150 Feet	175 Feet	200 Feet			
18	23	25	27	. . .	. . .	. . .	. . .	. . .	. . .	. . .	. . .	16	.97	1.77
21	35	38	41	. . .	. . .	. . .	. . .	. . .	. . .	. . .	. . .	19	1.47	2.41
24	49	54	58	62	. . .	. . .	. . .	. . .	. . .	. . .	. . .	22	2.08	3.14
27	65	72	78	83	87	. . .	. . .	. . .	. . .	. . .	. . .	24	2.78	3.98
30	84	92	100	107	113	119	. . .	. . .	. . .	. . .	. . .	27	3.58	4.91
33	105	115	125	133	141	149	. . .	. . .	. . .	. . .	. . .	30	4.48	5.94
36	128	141	152	163	173	182	191	. . .	. . .	. . .	. . .	32	5.47	7.07
39	154	168	183	196	208	219	229	. . .	. . .	. . .	. . .	35	6.57	8.30
42	182	200	216	231	245	258	271	288	. . .	. . .	. . .	38	7.76	9.62
48	. . .	269	290	311	330	348	365	389	. . .	. . .	. . .	43	10.44	12.57
54	. . .	348	376	402	427	449	472	503	551	. . .	. . .	48	13.51	15.90
60	. . .	436	471	503	536	565	593	632	692	748	. . .	54	16.98	19.64
66	. . .	. . .	579	620	658	694	728	776	849	918	981	59	20.83	23.76
72	. . .	. . .	698	746	792	835	876	934	1,023	1,105	1,181	64	25.08	28.27
78	. . .	. . .	. . .	885	949	990	1,038	1,107	1,212	1,310	1,400	70	29.73	33.18
84	. . .	. . .	. . .	1,035	1,098	1,157	1,214	1,294	1,418	1,531	1,637	75	34.76	38.48
90	. . .	. . .	. . .	. . .	1,269	1,338	1,403	1,496	1,639	1,770	1,893	80	40.19	44.18
96	. . .	. . .	. . .	. . .	. . .	1,532	1,606	1,712	1,876	2,027	2,167	86	46.01	50.27
100	. . .	. . .	. . .	. . .	. . .	. . .	1,760	1,865	2,043	2,197	2,359	89	50.11	54.54
104	. . .	. . .	. . .	. . .	. . .	. . .	1,899	2,024	2,218	2,395	2,560	93	54.39	59.00
108	. . .	. . .	. . .	. . .	. . .	. . .	2,051	2,190	2,399	2,591	2,770	96	58.83	63.62
112	. . .	. . .	. . .	. . .	. . .	. . .	. . .	2,323	2,588	2,795	2,983	100	63.46	68.42
118	. . .	. . .	. . .	. . .	. . .	. . .	. . .	2,632	2,883	3,114	3,339	105	70.71	75.94
120	. . .	. . .	. . .	. . .	. . .	. . .	. . .	2,725	2,986	3,225	3,447	107	73.22	78.54
124	. . .	. . .	. . .	. . .	. . .	. . .	. . .	2,915	3,193	3,449	3,687	110	78.31	83.86
130	. . .	. . .	. . .	. . .	. . .	. . .	. . .	3,165	3,467	3,745	4,004	116	85.04	90.76

surface offers less resistance to the passage of the gases, and there is not the likelihood of eddies being formed. There is much difference of opinion among engineers as to whether a stack should be narrower toward the top or increased in size. The practice is to taper a stack toward the top, this being done more on account of the necessity for increasing its stability than because of the draft. Some stacks have been built, however, with a larger inside diameter at the top than at the bottom, with the idea of providing a greater sectional area for the passage of the gases as their velocity is decreased. The capacity of the stack for carrying off the products of combustion depends on the temperature of the inside gases as compared with the temperature of the outside air. The average temperature in stacks for power purposes ranges from 450° to 600° F., and, therefore, as there is little difference in the travel of gases in flues between these temperatures, Table I can safely be used in determining the diameter and height of stack for a given capacity of power plant.

In Table I, it will be observed that the capacity of the stack is given in horsepower, and in calculating this table it was considered that 5 pounds of coal were burned to develop 1 horsepower, this being a high figure with the present economical systems of power generation. Allowance has also been made, in this table, for the friction of the gases against the side walls of the stack, it being considered that a 2-inch layer of dead air exists between the stack lining and the gases.

65. Stability of Brick Chimneys.—In considering the stability of brick stacks, the overturning moment due to the wind must not exceed the resisting moment of the stack to overturning about the base. For instance, referring to Fig. 31, the pressure p due to the wind acts with the lever arm x about the base of the stack, tending to overturn it. The stack, or chimney, resists this overturning moment with its weight w, acting through a lever arm y; if these two moments are equal, the stack can be considered safe under the conditions considered, though it is better to have some factor of safety, 2 usually being sufficient. An easy formula by which to determine whether a stack is stable or not, is as follows:

$$w = \frac{h^2 \times dc}{b}$$

in which w = weight of stack, in pounds;
h = height of stack, in feet;
d = mean diameter of stack, in feet;
c = constant;
b = width of base.

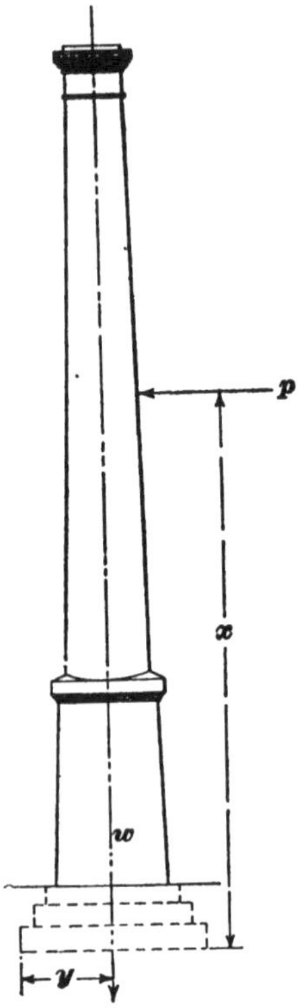

FIG. 31

The constant c varies with the shape of the stack. For a square stack, when the wind is blowing at hurricane violence, 56 is used; for an octagonal stack, 35; and for a round stack, 28.

To demonstrate this formula, consider a square chimney having an average breadth of 8 feet and a width at base of 10 feet, the stack being 100 feet high. The problem is, therefore, to find what the weight of the stack must be in order to resist the greatest wind pressure likely to occur. By substitution, in the formula,

$$w = \frac{100 \times 100 \times 8 \times 56}{10} = 448{,}000 \text{ pounds}$$

With brickwork weighing about 120 pounds per cubic foot, the chimney in question must therefore have an average thickness of somewhat more than 13 inches.

66. A good rule to follow in designing brick stacks is to make the base at least one-tenth of the height. For stacks under 5 feet in diameter, the walls for the first 25 feet from the top may be 8 inches, increased $4\frac{1}{2}$ inches for each additional 25 feet from the top. If the stack is more than 5 feet in diameter, the thickness at the top should be $1\frac{1}{2}$ bricks, or

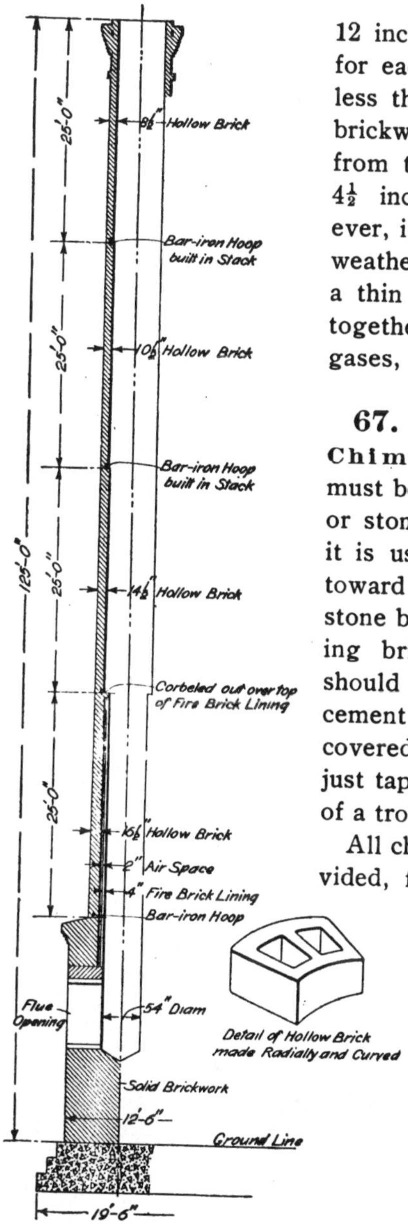

FIG. 32

12 inches, with a 4½-inch increase for each 25 feet. If the stack is less than 3 feet in diameter, the brickwork for the first 10 feet from the top may be as little as 4½ inches; this thickness, however, is not recommended, as the weather is likely to penetrate such a thin wall, and sooner or later, together with the exposure to the gases, destroy the brickwork.

67. Construction of Brick Chimneys.—All brick stacks must be provided with a cast-iron or stone coping at the top, and it is usually well to tie them in toward the base with good heavy stone band courses. In constructing brick stacks, the brickwork should be laid up in lime-and-cement mortar, and the bricks well covered and slid in place, not just tapped or hit with the handle of a trowel.

All chimneys should also be provided, for a distance of at least one-third of their height from the base, with a fire-brick lining, laid up in fireclay, and at the bottom of this lining, where the flues from the boiler enter the stack, cast-iron cleaning doors and frame should be provided for removing soot that will accumulate and drop down. A good example of a brick stack is given in Fig. 32; this stack has a capacity of 500 horsepower, and is sufficiently stable to resist any wind pressure.

FIRE-PROTECTION OF MILL BUILDINGS

SPRINKLER SYSTEM

68. Sprinkler Tanks.—In the large cities, where fire risks are great, and where nearly all the buildings and their contents are protected by insurance, the owners of the buildings are subjected to the rules and regulations of the Underwriters, or Associations of Insurance Companies. These Underwriters from time to time pass regulations insisting on certain further precautions and protection against fire, such as the installation of sprinkler systems, stand pipes for hose attachment for each floor, etc.

As the available city pressure or water supply of the municipality may be limited, or uncertain, or the pressure too low for a high building, it is sometimes necessary to place water tanks of from 10,000 to 30,000 gallons capacity in towers on the roofs of factories, and in the design of new factories provision is usually made for three tanks.

In designing a building, these tanks are located at such a point that their support is insured by the walls beneath, and the most convenient place is found to be over the stair tower or adjacent to it. As 1 gallon of water, together with the tank containing it, has a unit weight of 8 pounds, a 30,000-gallon tank complete will weigh in the neighborhood of 240,000 pounds, which must be supported on the walls and by means of iron beams.

The architect, besides providing adequate support for these tanks, must so design the tanks as to secure them against bursting, which would lead to serious consequences. For durability, sprinkler or fire-protection tanks are made of either cypress or cedar from 2 to 3 inches in thickness. They are usually in the shape of a truncated cone, and the bottom of the tank is required to be at least 20 feet above the highest point of the top story.

The important feature in the design of such tanks is to see that they are properly braced with hoops, and it is usual to specify that no hoop shall be subjected to a unit stress of more than 12,000 pounds for iron and 16,000 pounds for steel. These hoops are made from $\frac{3}{4}$-inch to 1-inch round iron, not less than the former, and the required strength is obtained by spacing them closer together at the bottom and farther apart toward the top. They are held together with adjustable clamps, as indicated in Fig. 33, and by the use of such clamps they may be readily tightened. The bottom hoops of the tank are subjected to great stress, and it is good

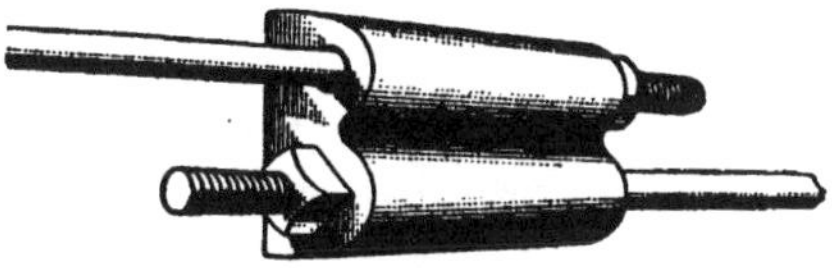

FIG. 33

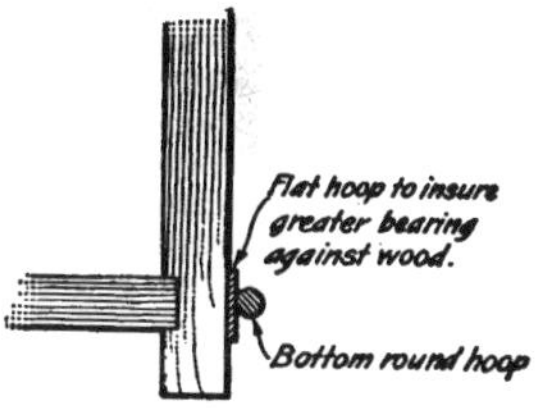

FIG. 34

practice for these hoops to bear against a flat iron hoop, as indicated in Fig. 34. By this construction much greater bearing is provided on the wood, and the round iron is prevented from cutting into the staves of the tank. In some instances flat iron hoops are used altogether, but it is considered better to use round iron hoops, from the fact that they are not likely to corrode through as rapidly as the thin flat iron.

69. Proportioning the Hoops.—The principal element of engineering entering into the design of large wooden water tanks consists in the proportioning of the hoops, and Table II will be found convenient in determining the hoops required for any size of tank.

70. In order to determine the number of hoops of a certain size required for any span of 12 inches at a point any distance from the water-line, the following formula may be used:

$$N = \frac{2.6\,d\,H}{S},$$

in which N = number of hoops required in 1 foot of height of tank;

d = diameter of tank, in inches;

H = height of water-line from center of space under consideration, in feet;

S = actual safe strength, in pounds, of hoops assumed to be used.

This last value may be found from Table II.

TABLE II

SAFE STRENGTH OF ROUND TANK HOOPS

Diameter Inch	Steel Pounds	Wrought Iron Pounds
$\frac{5}{8}$	3,232	2,424
$\frac{3}{4}$	4,832	3,624
$\frac{7}{8}$	6,720	5,040
1	8,800	6,600

71. To illustrate the foregoing, assume that it is desired to find what will be the spacing of $\frac{3}{4}$-inch iron hoops at the bottom of a tank 12 feet in diameter, in which the water-line is 16 feet from the middle of the section under consideration. Applying the formula in Art. **70,** using in conjunction therewith Table II, it is found that

$$N = \frac{2.6 \times 144 \times 16}{3{,}624} = 1.7$$

This result, 1.7, is the number of hoops required in 12 inches of height from the bottom of the tank, and would indicate that the hoops should be spaced about 7 inches from center to center, for 12 inches divided by 1.7 gives approximately 7 inches, the pitch of the hoops. This process should be repeated for different points throughout the height of the tank, and from the results the tank may be designed.

72. In the installation of sprinkler tanks, it must be observed that they are placed some distance above the highest point of the top floor, the distance usually required by the Underwriters being 20 feet, if it is possible of attainment. The tank should always be roofed, have a ladder from the roof of the building to its top, and a steam pipe inside to prevent the water from freezing in winter. This pipe is furnished with a check-valve to prevent the water in the tank from siphoning.

EXAMPLE FOR PRACTICE

What should be the spacing of the $\frac{3}{4}$-inch round wrought-iron hoops on a tank 10 feet in diameter and 12 feet high at a distance of 6 feet from the water-line? Ans. 23.2 in.

---

73. Automatic Sprinkler System.—The sprinkler system as now installed for protection against fire in the interior of a building consists essentially of piping connected to a gravity tank and extending over the entire ceiling by means of mains and branches. There is located on the ends of the branches automatic valves or stops, which are collapsed or opened by the melting of a fuse or solder at a temperature more than is likely to exist in the room at any time and still below that which would be created by an incipient fire.

74. The underlying principles of automatic sprinkler systems as stated by the Underwriters are as follows:

1. Buildings must be open in construction, free from concealed spaces, or places where water thrown from sprinklers cannot penetrate.
2. Sprinklers to be so located that their distribution will cover all parts of the premises.
3. Sprinkler piping to be of sufficient capacity and to have water under pressure in same at all times, except in case of a system where freezing is likely to occur, where an air lock is used.

4. An automatic supply of water of sufficient quantity and pressure available at all times.

5. Systematic, thorough, and intelligent care and inspection of the system.

75. Fireproof Windows.—It is frequently necessary, and in many cases required by law, and especially recommended by the Underwriters, to provide fireproof window frames and sash in walls exposed to great fire risk, or where it is necessary to admit light into elevator shafts or fire-towers. To meet this demand, several forms of metallic window frames and sashes have been evolved, and these sashes when intended as a fire-retarder are always glazed with wired glass.

76. Wired Glass.—The wire glass now in common use consists of heavy glass plate with wire mesh embedded in it. This glass is obtainable in polished, ribbed, prism, or mazed form, as shown in Fig. 35 (*a*), (*b*), (*c*), and (*d*), respectively. The plain glass, Fig. 35 (*a*), is used where the light is ample, and where it is desired for the occupants to see through the windows. The ribbed is employed usually in factories, and the ribs are generally run in a horizontal direction, so as to throw the light toward the ceiling and floor, thus diffusing it throughout the building. The prism glass is also employed in order to secure a greater diffusion of the light than is possible with the plain or factory ribbed glass, while the mazed glass finds favor where it is necessary to employ an obscured sash, which will still admit plenty of light and present a good appearance but yet cannot be seen through.

The glass used in metallic frames should not be less than $\frac{1}{4}$ inch, or, if polished, $\frac{5}{16}$ inch, and the embedded wire should not have a mesh larger than 1 inch and should not be less in size than No. 22 Brown & Sharpe wire gauge, which is the standard used in America.

77. Design of Sash.—In designing a sash for fire-retarder frames, it is necessary, in order to comply with the Underwriters rules and regulations, to observe that no single light exceeds 24 in. × 30 in. The metallic frames are generally constructed of No. 22 galvanized steel, while the sash are made of a lighter weight, generally No. 24. In unusual localities, where the frames are likely to be subjected to the

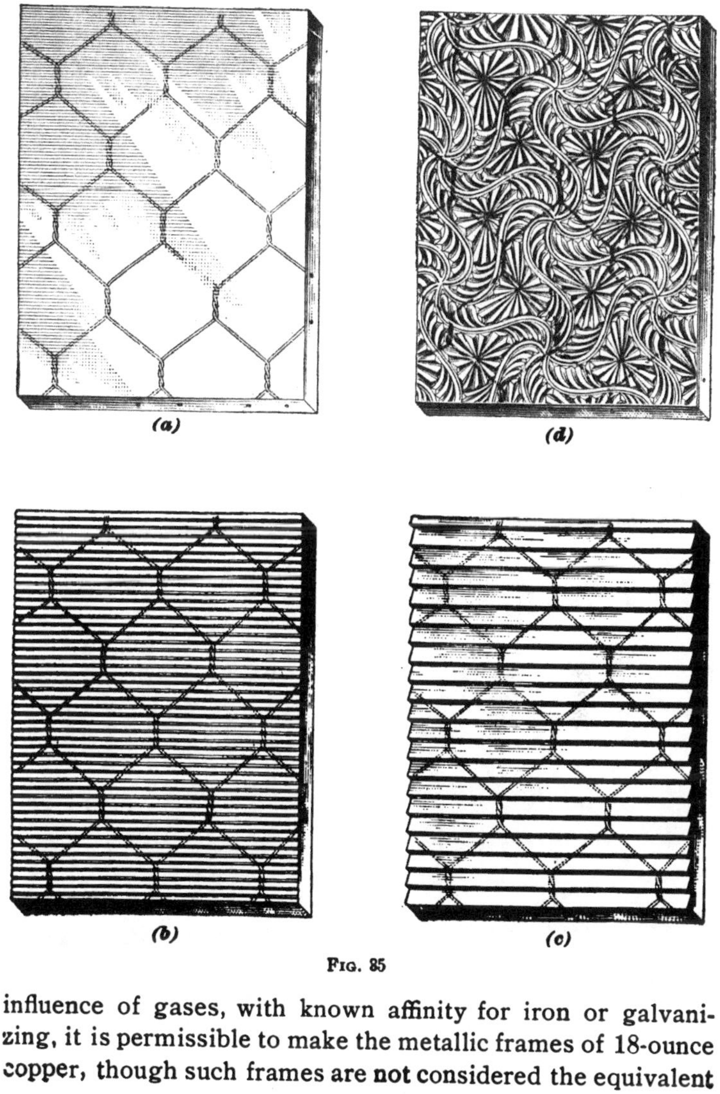

(a) (d)

(b) (c)

Fig. 85

influence of gases, with known affinity for iron or galvanizing, it is permissible to make the metallic frames of 18-ounce copper, though such frames are not considered the equivalent of an iron frame as a fire-retarder, and such frames should never be used in elevator, vent shafts, or fire-retarder partitions that are liable to intense internal fires.

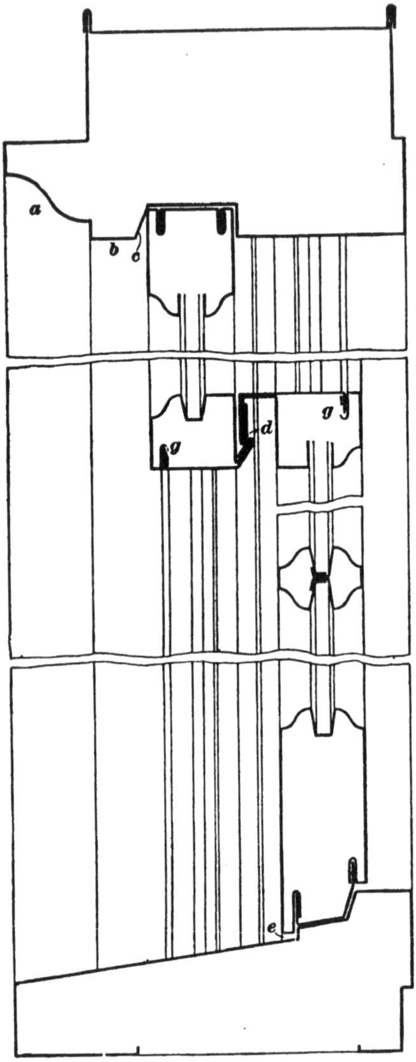

FIG. 36

In order to better explain the construction of the commercial frames, Fig. 36 is given, which illustrates one of the best frames in the market. In the figure, a vertical cross-section through the window-head sill and parting rail is illustrated. It will be observed that these frames can be neatly framed with architrave mold and stop, as designated at *a* and *b*. It will also be observed that the head for the top sash is beveled, as indicated at *c*, so that a tight joint is insured by the edge of the sash coming in contact with the bevel, and thus compelling a close connection. The parting rails are also constructed with a straight piece entering on a bevel *d*, so that at this point a tight joint is also secured. By the several offsets in the sill, wind and rain stops are provided, as indicated at *e*. Sashes constructed in this manner can be made to slide freely, noiselessly, and be made tight against weather and wind, as well as being secured against annoying clattering, or rattling. The sills of metallic frames are generally filled with cement, and sometimes the heads are similarly made solid. Any unusually large surfaces, like that which would occur between twin or triple windows, in the mullion, are securely braced inside with galvanized sheet iron or bar iron.

78. In the construction of metallic sash, solder is never used for holding the parts together, for all parts must be either lock-seamed or riveted, the lock seams being illustrated at *g*, *g*, Fig. 36. Soldering may be used only to fill up the joints. The objection to a joint that is only soldered and not lock-seamed is that in a severe fire when the window is subjected to an intense heat, the joint is apt to open by the solder melting out. When the joint opens, flames may go through and the fire-stop will thus be soon destroyed.

In designing the frames, they should have at least a 4-inch lap on the brick reveal on the sides and head, and it is not uncommon to wind-stop the sill by extending upwards a piece of galvanized sheet iron. While such windows as those described will act as a fire-retarder and prevent flames from reaching apartments that they protect, even in cases of severe conflagrations, nevertheless the glass radiates considerable heat, and inflammable goods should not be stored too close to such windows. Neither is it particularly desirable to have window shades secured to the frames of metallic windows. Where the goods in a building are particularly inflammable, the liability to pile them too close to the sash should be entirely eliminated by using window guards, which would maintain such merchandise at a distance of 3 or 4 feet from the window.

79. Fire-Doors and Frames.—There is no more important feature in the design of a mill building than the tin-lined fire-doors and their attachment to the jambs. Every fault in their construction, as viewed by the Underwriters, is likely to cost the owner additional insurance.

All tin-lined doors, when one door is used, should be made of three thicknesses of tongued-and-grooved planking, laid up and down and horizontally, and clinched-nailed, as illus-

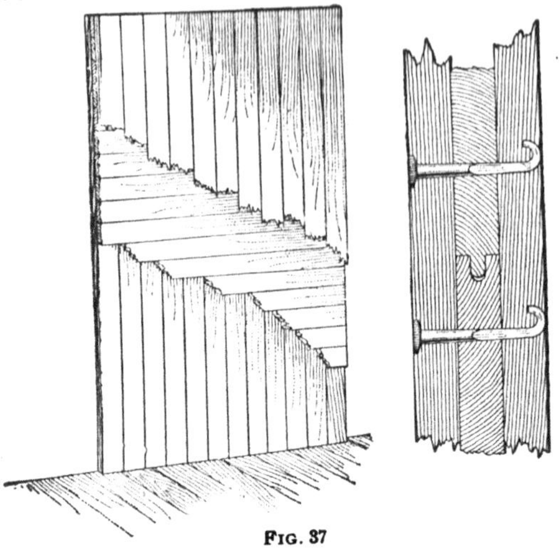

FIG. 37

FIG. 38

trated in Fig. 37. The tin lining on these doors must be of IC tin, put together with locked seams, secretly nailed, and presenting the appearance designated in Fig. 38.

80. The sills of openings covered with tin-lined doors must always project under the door, so that there is no danger of burning through the floor and thus communicating to the space protected by this entrance. The several constructions of sills most commonly used are illustrated in Fig. 39.

81. Sliding doors should be hung with anti-friction adjustable hangers. That is, the wheel of the hanger should have roller bearings for the axle, and there should be some means of adjusting the height of the door above the threshold by

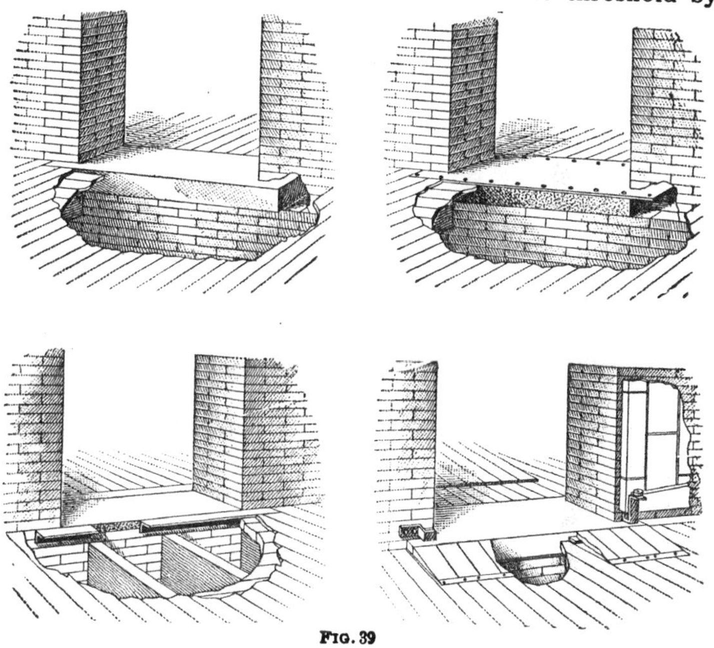

FIG. 39

means of the hanger. The track for sliding doors should be placed on a slant toward the opening, so that the door will automatically close. Where it is desired to have the door open, it may be held back by means of a chord, fusible link, and counterweight.

82. All folding doors should be heavily strap-hinged, and secured to the jambs with iron-hanging stiles and hinge eyes with through bolts, as shown at *a*, Fig. 40.

Care must always be taken that any through bolts that go through brick walls near door openings, as the bolts shown at *a*, Fig. 40, be far enough away from the jambs so that

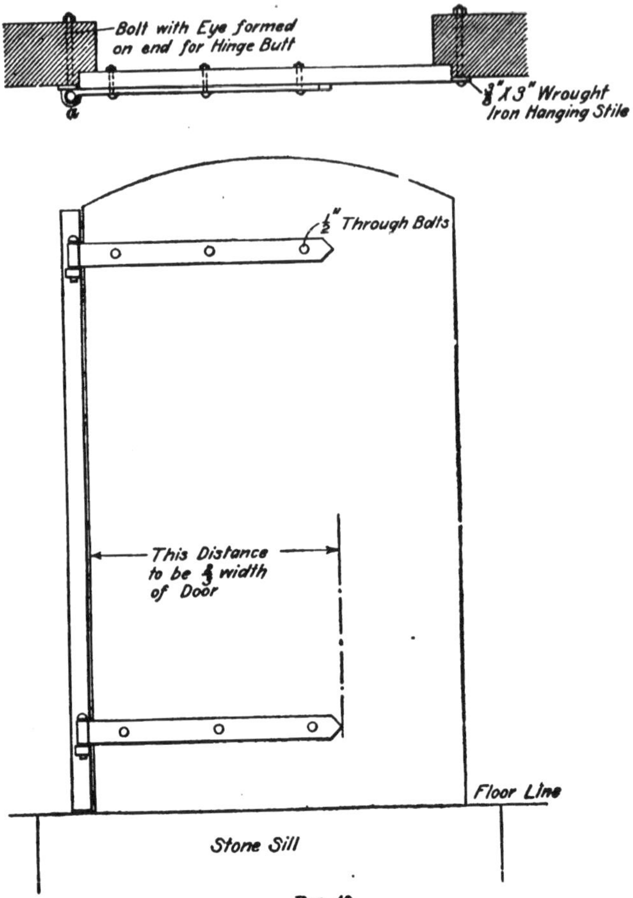

FIG. 40

there will be no danger of the bolts pulling through when put under strain. It is always better to build these bolts in the wall as the work progresses than to drill holes and put them in afterwards.

Spotted something that isn't quite right?

Let us know and we'll fix it for future reprints.
And as a thank you for your diligence, we'll
send you a free Sothis book of your choice.

publisher@sothis.press

This page intentionally left blank

Except of course for that line above. And this one.
And whilst we have your attention, perhaps you'd
like to follow us on Instagram @sothis.press. Go on.

Made in the USA
Monee, IL
05 November 2026

67807977R00097